NEUROSCIENCE RESEARCH PROGRESS

CHEMOSENSORY SENSORS AND SYSTEMS

EVOLUTIONARY SIGNIFICANCE, BIOLOGICAL EFFECTS AND NEW INSIGHTS

NEUROSCIENCE RESEARCH PROGRESS

Additional books in this series can be found on Nova's website
under the Series tab.

Additional e-books in this series can be found on Nova's website
under the eBooks tab.

NEUROSCIENCE RESEARCH PROGRESS

CHEMOSENSORY SENSORS AND SYSTEMS

EVOLUTIONARY SIGNIFICANCE, BIOLOGICAL EFFECTS AND NEW INSIGHTS

MAXIMILIAN BRANDT
EDITOR

nova
biomedical
New York

Library of Congress Cataloging-in-Publication Data

ISBN: 978-1-53612-406-4

Published by Nova Science Publishers, Inc. † New York

CONTENTS

PREFACE

Chemical sensor is a self-contained analytical device that can provide information about the chemical composition of its environment, that is, a liquid or gas phase. Chapter One attempts to integrate past and current models of oxygen sensing and review their evolutionary importance as they relate to adaptations within the oxygen transport cascade. Chapter Two reports on improved electrochemical performances such as higher sensitivity, limit of detection ; linear dynamic range, response time, and long-term stability towards hydrazine that were achieved using the modified chemical sensor. Chapter Three is a précis of modern olfactometer techniques across research disciplines in order to collate their growing complexity and utility.

Chapter 1– The oxidation of earth's atmosphere radically shaped the adaptive trajectory of multicellular life. The advent of photosynthetic oxygen production ~2.7 billion years ago brought with it the rapid and permanent oxygenation of the atmosphere and ocean. While the toxicity of oxygen and reactive oxygen species (ROS) presented a crisis to the existing anaerobic biota, it also exerted a significant selection pressure favouring anatomical and cellular infrastructure that enabled organisms to not only tolerate hyperoxia, but allow them to exploit new oxidative processes that could generate metabolic energy. On an individual level, physiological oxygen conditions are highly variable throughout an organism's lifecycle. For instance, hypoxemia can be observed naturally in healthy developing embryos where it acts as an important micro-environmental signal in coordinating the differentiation of progenitor cell s. Similarly, healthy adults experience regional reductions in oxygen availability due to exercise -induced increases in metabolism, which stimulate the systemic pathways vital to maintaining an adequate blood supply to muscles. Conversely, hypoxia is a central component of a wide range of pathophysiological processes, including heart attack and stroke, chronic lung disease s, cancer, and many neurodegenerative disorders. Taken together, these multitudinous and variable responses to changes in oxygen availability have fuelled considerable interest in the biological mechanisms underlying physiological oxygen homeostasis.

Chapter 2– Nanocomposites of strontium oxide nanoparticles decorated on carbon nanotube (SrO.CNT NCs) were prepared in basic medium using a wet-chemical method. The SrO.CNT NCs were investigated using Fourier transform infra-red spectroscopy (FTIR), Ultraviolet- visible spectroscopy (UV-Vis.); Field emission scanning electron microscopy (FESEM) equipped Electron dispersive spectroscopy (EDS), Energy dispersive X-ray spectroscopy (XPS), and X-ray powder diffraction (XRD) techniques. A selective hydrazine sensor onto a flat glassy carbon electrode (GCE) was fabricated with a thin-layer of NCs. Improved electrochemical performances such as higher sensitivity, limit of detection ; linear dynamic range, response time, and long-term stability towards hydrazine were achieved using the modified chemical sensor. The calibration curve was found linear over a wide range of hydrazine concentration. The sensitivity and detection limit were calculated based on slope/surface area of GCE, and signal to noise ratio ~ 3 respectively. It is the first time report regarding hydrazine detection using current *vs* voltage (I-V) process by means of SrO.CNT NCs modified GCE. The synthesis of SrO.CNT NCs using a wet-chemical procedure is a good way of establishing chemical sensor based on nano-materials (NMs) for toxic elements. It is also commenced a potential development of hydrazine sensor using SrO.CNT NCs by a simplistic and dependable I-V technique for the prospective appliance of harmful agents detection in environmental and health care area. Finally, the effectiveness and efficiency of the proposed selective chemical sensors could utilized significantly and effectively for the detection of toxic hazardous chemicals for safety green environment and healthcare fields in broad scales.

Chapter 3– The olfactory sense is difficult to measure. Measurement of olfaction covers at least four separate paradigms, all of which require separate methodologies to analyse, and all of which are subject to a litany of known and unknown variables. As a result, devices used to measure olfaction, olfactometers, have progressively developed variations and sophistication. Despite their comparative infancy, olfactometers have been established across a broad spectrum of disciplines; primarily psychology, neuroscience, biology, food science, and environmental engineering. While these separate disciplines have all shown a need for olfactometer technology, there is also very little cross-disciplinary information sharing. This chapter is a précis of modern olfactometer techniques across research disciplines in order to collate their growing complexity and utility.

In: Chemosensory Sensors and Systems ISBN: 978-1-53612-406-4
Editor: Maximilian Brandt

Chapter 1

OXYGEN SENSING AND TRANSCRIPTIONAL REGULATION OF ADAPTIVE HYPOXIC RESPONSES

Lisa G. Borecky and Matthew E. Pamenter, PhD*
Department of Biology, University of Ottawa,
Ottawa, ON, Canada
Brain and Mind Research Institute, University of Ottawa,
Ottawa, ON, Canada

INTRODUCTION

The oxidation of earth's atmosphere radically shaped the adaptive trajectory of multicellular life. The advent of photosynthetic oxygen production ~2.7 billion years ago brought with it the rapid and permanent oxygenation of the atmosphere and ocean [1]. While the toxicity of oxygen and reactive oxygen species (ROS) presented a crisis to the existing anaerobic biota, it also exerted a significant selection pressure favouring

* Author's Email: mpamenter@uottawa.ca.

anatomical and cellular infrastructure that enabled organisms to not only tolerate hyperoxia, but allow them to exploit new oxidative processes that could generate metabolic energy [2]. On an individual level, physiological oxygen conditions are highly variable throughout an organism's lifecycle. For instance, hypoxemia can be observed naturally in healthy developing embryos where it acts as an important micro-environmental signal in coordinating the differentiation of progenitor cells. Similarly, healthy adults experience regional reductions in oxygen availability due to exercise-induced increases in metabolism, which stimulate the systemic pathways vital to maintaining an adequate blood supply to muscles. Conversely, hypoxia is a central component of a wide range of pathophysiological processes, including heart attack and stroke, chronic lung diseases, cancer, and many neurodegenerative disorders. Taken together, these multitudinous and variable responses to changes in oxygen availability have fuelled considerable interest in the biological mechanisms underlying physiological oxygen homeostasis.

The immediate cellular impact of low oxygen manifests at the mitochondrial level, where oxygen participates in the mitochondrial respiratory chain, which couples the combustion of organic compounds to the production of adenosine triphosphate (ATP) in a process known as aerobic respiration [3, 4]. Oxygen acts as the terminal electron acceptor at complex IV (cytochrome c oxidase; COX) located within the inner mitochondrial membrane, where its reduction is linked to the efflux of H^+ ions (protons) out of the mitochondrial matrix. This creates an electrochemical H^+ gradient large enough to drive protons through the ATP-synthase, which powers the phosphorylation of ADP to form ATP. From start to finish, the aerobic combustion of glucose yields ~ 38 molecules of ATP. The high energetic output associated with the reduction of oxygen—a 19-fold increase over simple glycolysis employed by anaerobic organisms—supported the ascension of large and complex organisms. During the course of evolution, however, many organisms became completely reliant on oxygen for survival [2, 5].

The potentially lethal consequences of oxygen deprivation have driven the evolution of physiological mechanisms that allow complex multicellular systems to respond to changing oxygen availability and preserve cellular integrity within an organism's ecophysiology. As organisms become larger and more active, oxygen transport by simple diffusion becomes limiting. As a result, respiratory and circulatory systems within the animal kingdom exhibit considerable variability in terms of size, organization, and capacity to detect and respond to changes in oxygen availability at either a local or systemic level. For example, insects possess a dense network of transverse and longitudinal trachea which continuously deliver atmospheric oxygen into the insects' tissues. This direct contact between air and respiring mitochondria affords insects their highly active lifestyles and allows them to exhibit behaviours associated with high metabolic expenditure such as flight [6-8]. On the other hand, vertebrates possess bodily fluids into which oxygen must be diluted after passing across a respiratory diffusion interface. In contrast to the direct diffusion of oxygen in insects, this strategy of oxygen transport is inefficient; however, increasing complexity and proportional growth of the respiratory and cardiovascular system s eliminates the size restriction imposed on insects. Furthermore, much of this inefficiency is mitigated by haemoglobins—a versatile class of proteins found in red blood cells that is capable of reversibly binding oxygen, and which transports oxygen from the lungs to respiring tissues [9, 10]. For example, in humans haemoglobins confer a ~70-fold increase in blood oxygen saturation and thus dramatically increases the oxygen transport efficiency of an individual [10].

Beyond increasing complexity of larger animals, naturally occurring oxygen -limited environments have propelled the evolution of individual capacities to tolerate low oxygen environments to incredible heights. This process provides researchers with several model organisms that are capable of surviving in hypoxia or anoxia for weeks to months. Indeed, animals exhibit adaptations throughout the oxygen transport cascade that help to ameliorate the deleterious effects of systemic hypoxia. For example, changes in ventilation, cardiac perfusion, blood -oxygen carrying capacity, or diffusibility of blood from the lungs to the blood and from the blood to the tissues are all modified in some organisms to help ameliorate the effects of environmental hypoxia [11].

The success of adaptive hypoxic responses is predicated on the ability to sense and communicate changes in oxygen availability in order to initiate cardiorespiratory, metabolic, and/or behavioural adaptations that serve to either increase oxygen delivery or decrease metabolic expenditure. Ultimately, sensor responses depend on their inherent affinities for oxygen. The wide range of oxygen tensions within the oxygen transport cascade may in part reflect differences in the preset values of cells, tissues, and/or organ systems at which adaptive responses are initiated. In this chapter we attempt to integrate past and current models of oxygen sensing and review their evolutionary importance as they relate to adaptations within the oxygen transport cascade. Foundational theories of oxygen sensing at the cellular and organ level are presented, ranging from direct oxygen-sensing mechanisms to secondary messengers, and the unique role for the mitochondria as the cellular powerhouse. Special emphasis is placed on a highly conserved oxygen-sensitive system of gene regulation mediated through the induction of hypoxia -inducible transcription factors (HIF). The regulatory components of this ubiquitous hypoxic response pathway can be found in animals as simple as fruit flies and as complex as humans, highlighting HIF as an ideal target to mediate and integrate oxygen-sensing systems across the animal kingdom.

Peripheral Chemoreflexes: From Fish to Mammals

Peripheral oxygen chemoreceptors that drive changes in cardiac and ventilatory activity in response to alterations in arterial oxygen, carbon dioxide, or pH have been identified across animal taxa. In fish, chemosensory neuroepithelial cells (NECs) are found in the skin of developing embryos, adult gills, in the respiratory gas bladder (e.g., the longnose gar) [12], and/or in air-breathing organs [13]. NECs have been identified in the gills of every fish species studied to date and their role in initiating respiratory responses to changes in oxygen levels is well-established [14-16]. In adult fish, gill NECs have been primarily documented in the distal half of the primary gill filaments where they are situated near the efferent filament artery [17-20]. As such they are exposed to both the external environment and the arterial blood supply making them well-suited to detect changes in respiratory gases [18, 19]. The gills are innervated by cranial nerves IX and X, also referred to as the branchial nerves [21]. Zebrafish and goldfish also possess nerve fibres that stem from the nerve bundle located between the central venous sinus and the efferent filament artery and which lie adjacent to lamellar NECs [19, 20]. The close proximity of NECs to nerve fibres suggests changes in oxygen availability are communicated via neurotransmitter release.

Neuroepithelial cells were first documented in fish and elasmobranchs by Dunel-Erb et al., (1982), who observed morphological similarities between cells within the gill and previously described neuroepithelial bodies (NEBs), which are found within the respiratory tract of mammals. However, it was Milsom's group who first demonstrated hypoxia - mediated increases in afferent sensory activity in the isolated first gill arch of yellow-fin tuna [22-24]. Similar responses were reported from isolated gill arches of the rainbow trout from Burleson and Milsom (1992). Indeed, it is now appreciated that gill NECs share a similar embryonic origin [25], structural characteristics [19, 26], patterns of afferent nerve activity [22-24], and transduction mechanisms (Jonz et al., 2004). When stimulated by hypoxic waters, hypercapnia, high levels of ammonia, or extracellular acidosis, fish will exhibit profound physiological responses. For example, when exposed to low oxygen tensions, fish will typically augment water propulsion over their gills either by rapid and reversible increases in both the amplitude and frequency of breaths (non-ram ventilators), and/or changes in swimming activity (e.g., move to more oxygenated waters) [27, 28]. Fish exposed to acute hypoxia will also experience reflex changes in heart rate (bradycardia), and elevated gill vascular resistance [29]. The combined effect is an increase in oxygen delivery to respiring tissues necessary to maintain cellular viability in low oxygen conditions.

Interestingly, profound inter-species differences exist between teleosts in the magnitude of their cardiorespiratory responses to hypoxia, which may relate to variability in the position, orientation, and distribution of oxygen chemoreceptors [30-34]. For example, the tambaqui (*Colossoma macropomum*), a neotropical freshwater fish, exhibits a 63% reduction in heart rate and a 57% increase in respiratory rate when exposed to hypoxic conditions [34]. Tambaqui possess oxygen chemoreceptors along both the internal and external gill surfaces, as well as extrabranchial chemoreceptors, which contribute seemingly equally to ventilation and cardiac performance [32, 34]. On the other hand, greater increases in breathing frequency (~90%) but only moderate bradycardia (15% reduction in hear rate) are observed in the bowfin (*Amia calva*)—a temperate demersal freshwater fish [35]. In contrast to the tambaqui, oxygen chemoreceptors within the gill of the bowfin appear not to contribute to the cardiac response, but aggregations of chemoreceptors orientated internally and externally regulate hypoxic ventilatory responses [35].

Based on the high degree of similarity between the chemoreceptive cell characteristics of fish and other vertebrates, peripheral chemoreceptive sites in terrestrial vertebrates are believed to be evolutionarily descended from fish NECs [27, 36]. Among non-mammalian vertebrates, the process of respiratory control through chemoreceptive input is best characterized in fish; however, chemoreceptors homologous to fish NECs and mammalian carotid bodies have been observed in amphibians [37-40], birds [41, 42], and most recently, reptiles [43, 44]. In larval amphibians, chemoreception begins in both the skin and gills, whereas, upon metamorphosis, anatomical and physiological reorganization of the gill arches mediates the transition from gill to lung breathing to support life on land [27]. In adult amphibians, the peripheral chemoreceptors that control lung ventilation and cardiac responses include the carotid labyrinth, the pulmocutaneous artery, and the aortic trunk (reviewed in [38]. The amphibia n carotid labyrinth developed from the first extant gill arch, and rests on the bifurcation of the carotid arteries where it acts as the primary site of chemoreception in the adult [27]. A study by Jia and Burggren (1997) supports the role of the first gill arch in mediating rapid ventilatory responses to hypoxia. These authors observed that an increase in ventilation associated with exposure to sodium cyanide was eliminated or greatly delayed following bilateral ablation of gill arch I [45].

More recently, Reichert et al., (2015) provided direct evidence for the existence of a chemoreceptor population at the first carotid bifurcation in the central vascular area in the lizard *Tupinambis merianae*. Their results also demonstrate these chemoreceptor cells share similar properties to the NECs of fish and the carotid body of mammals [43]. In birds, the homologues of these chemoreceptive sites are not positioned at the carotid body bifurcation but instead retreat backwards toward the central cardiovascular area close to the parathyroid and thyroid gland s [27, 41]. The external carotids then atrophy and the internal carotids (or common carotids) secondarily divide in the head region [46]. A unique group of peripheral receptors also exists in avian lungs called the intrapulmonary chemoreceptors, which are capable of responding rapidly to altered intrapulmonary carbon dioxide concentration in order to initiate reflex changes in ventilation [47, 48].

In mammals, chemoreceptive sites are typified by type I cells of the carotid body [49, 50], NEBs of the lung [51, 52], pulmonary arterial smooth muscle cells (PASMCs), and the aortic bodies of the aortic arch—the latter two of which may participate in regulating cardiovascular responses but contribute little to ventilatory drive [53]. Like amphibians, mammalian carotid bodies are situated at the bifurcation of the common carotid arties. NEBs are a group of richly innervated polymodal sensors dispersed within the airway epithelial of mammalian lungs. NEBs store amines (e.g., serotonin) and peptide neurotransmitters /neuromodulators, and most likely release these in response to hypoxia [54]. Their sensory function has been incompletely characterized due to their low density and widespread distribution in lung tissue; however, studies involving electrophysiological patch clamping of isolated NEBs confirm they are excitable and possess voltage-activated Ca^{2+}, Na^{+}, and K^{+} currents similar to carotid body glomus cells [55]. Furthermore, studies have also reported that these K^{+} channels modulate neurotransmitter release upon exposure to hypoxic stimulus [52, 56, 57].

In general, oxygen -sensing and transduction pathways are highly conserved across aerobic species despite major anatomical differences [27]. For example, NECs are scattered throughout the gills as well as the oro-branchial cavity and their orientation and relative contribution to each component of physiological responses to environmental hypoxia in fishes can be highly variable between species. Mammals on the other hand typically have a dominant concentration of chemoreceptive tissue —the carotid body—which contributes to the regulation of ventilation. This divergence likely reflects the transition from aquatic respiration and constantly fluctuating oxygen conditions, to air-breathing in a stable atmosphere [27]. Mechanisms of oxygen sensing and signal transduction are best characterized in fish and mammals, and thus this chapter will focus on these model system s. Mechanisms in other species will also be discussed in the context of understanding mammalian systems and/or with regards to their participation in the evolution of oxygen sensing systems.

Carotid Body Oxygen Sensing and Transduction

The bulk of evidence now suggests all mammalian tissues sense and respond to low oxygen, albeit at varying thresholds. Carotid bodies, however, are characterized by a uniquely rapid increase in sensory activity within seconds of the onset of hypoxia. Additionally, physiological responses may be initiated by the carotid bodies at even modest levels of hypoxia (~80 mmHg) [58]. Once the low-oxygen signal is coupled with its transducer to induce membrane depolarization, neurotransmitter release is necessary for afferent nerve activation and subsequent autonomic changes in breathing and blood pressure. Indeed the carotid body possesses a large number of putative neurotransmitters, including biogenic amines (acetylcholine, dopamine, etc.), neuropeptides (substance P and endothelins), ATP, amino acids etc. [59]. Nitric oxide (NO) and carbon monoxide (CO) have also been categorized as unconventional neurotransmitters within the carotid body.

Parallel to other sensory pathways, the neurotransmitters expressed in the carotid body can either excite or inhibit afferent neural discharge. The previous notion that the communication of hypoxia exclusively involves excitatory transmitter(s) has been disproven. For example, there is a clear role for inhibitory dopaminergic signalling in the hypoxic response of the carotid body [60, 61]. Instead it has been suggested that inhibitory and excitatory neurotransmitters operate in tandem, comprising a push-pull mechanism to sustain carotid body excitation [58, 62]. The interactions between various neurotransmitters may play an important role in the plasticity of carotid body responses to chronic hypoxia, and this idea will be discussed below. For a more exhaustive review of carotid body neurotransmitters see [62-64].

Carotid bodies are small, paired, highly vascularised organs positioned at the bifurcation of the common carotid arteries. In mammals, the carotid body parenchyma is made up of glomus cells (also known as type I cells), which contain numerous neurotransmitters and lie in synaptic contact with afferent nerve fibres joining the glossopharyngeal and vagus nerve s [65]. Carotid bodies also possess sustentacular cells (type II cells) which are classically believed to provide the carotid body with structural support, but more recently have been suggested to participate in coordinating chemosensory transduction through interactions with other cells in the carotid body via 'gliotransmission' mediated by pannexin 1 channels [66, 67]. There is compelling evidence that Type I cells are the principal gauges of multiple arterial chemical inputs —including O_2, CO_2, pH, $[K^+]$, glucose and other osmolytes, temperature, and blood pressure —and are thus the true transducers of carotid body chemostimuli [68]. The activity of the carotid sinus nerve reflects changes in airway gas concentration—which is primarily affected by the adequacy of ventilation relative to metabolism [69, 70]. When stimulated by either environmental hypoxia or regional ischemia, glomus cells release excitatory neurotransmitters—the most notable being ATP, dopamine, and acetylcholine —that transmit information across the synaptic cleft to trigger an action potential which moves along adjacent terminals to the brainstem respiratory centre to produce effective cardiorespiratory responses [70, 71]. The afferent discharge of the carotid body is nonlinear, and produces a graded response curve with corresponding oxygen levels.

Neurotransmission in the carotid body is multifaceted and multiple mechanisms of signal transduction that couple decreases in arterial oxygen content to transmitter release have been proposed and tested (summarized in Table 1). Little, if any, compelling evidence exists supporting the idea that a single O_2 sensor is responsible for the entire stimulus -response characteristic of the carotid body. Instead, it is now generally appreciated that a variety of signals, sensors, and transmission mechanisms each with varying oxygen affinities and thresholds for hypoxia work in concert to engage protective hypoxic mechanisms. This redundancy in oxygen sensing mechanisms is indicative of the importance of maintaining cellular integrity during hypoxic stress.

Table 1. Proposed mechanisms of oxygen sensing and signal transduction

Sensor	Coupler	Effector	Downstream Target/Output
K^+ Channel	O_2	Direct/CB	Ventilation/Cardiac Output
	CO	HO-2/ CB	Ventilation/Cardiac Output
	H_2S	HO-2/ CB	Ventilation/Cardiac Output
	ATP	AMPK/CB	Ventilation/Cardiac Output
NOS	NO	K^+ Channel	Ventilation/Cardiac Output
Ca^{2+} Channel	O_2	Direct/CB	Vasodilation
Na^+ Channel	O_2	Direct/CB	Ventilation/Cardiac Output
Mitochondria	ROS	K^+ Channel	Ventilation/Cardiac Output
	ROS	Ca^{2+} channels	Ventilation/Cardiac Output
	ATP	AMPK	Protein Synthesis/Metabolic Depression
NADPH Oxidase	O_2	K^+ Channel	Ventilation/Cardiac Output
	ROS	PHDs/HIF-1	Multiple
Heme/Globins	O_2	Na/H, $K^+C\,l^-$ exchangers	pH/ loading and unloading efficiency
PHDs/AHDs	O_2	HIF-1	Multiple
Mitochondrial Cytochromes	ATP	K^+ Channel	Ventilation/Cardiac Output

Membrane Hypothesis: K^+ Channels as Oxygen Sensors

A unifying consensus underlying the oxygen sensing and transduction mechanisms of peripheral chemoreceptors is the modification of ion channels and subsequent changes in cellular gradients that control downstream signalling cascades and 2nd messenger systems critical to cellular viability. Historically, ion channels responsible for the influx of K^+ have been the focus of hypoxia -induced membrane depolarization. The inhibition of K^+ channels by reductions in oxygen leads to the activation of voltage-gated Ca^{2+} channels, resulting in Ca^{2+} influx [72]. The elevated intracellular Ca^{2+} in type I cells drives the excitation of afferent neurons following the release of species -specific neurotransmitters by chemoreceptor cells and (Figure 1).

The notion that type I cells in the carotid body possess oxygen - sensitive elements was first proposed by Lopez-Barneo et al., (1988) who demonstrated that type I cells were electrically excitable and that reductions in PO_2 induced reversible inhibition of big potassium (BK) channels (also called Maxi-K or Slo1 channels) in the glomus cell of rabbits [73]. Further support for this idea came with the application of the whole cell patch clamp technique to isolated type I cells *in vitro,* which further revealed that selective closure of K^+ channels in type I cells in fact induced depolarization, resulting in an increase in cytosolic $Ca^{2+,}$ neurotransmitter release, and the excitation of afferent sensory fibres [74-77]. Previous interpretations that K^+ channels possessed oxygen-sensitive, pore-forming subunits, came from the assertion that K^+ channel inhibition was membrane restricted and required no diffusible factors [75].

The regulation of K^+ channels is complex, and after three decades of research, the exact identity of the upstream sensing machinery and even the molecular signal which elicits inhibition of K^+ channel activity remains elusive. Following the discovery of oxygen -sensitive K^+ channels in type I cells, similar observations were made by a number of research groups [78, 79]. Multiple K^+ channel types and subtypes have now been identified and shown to be inhibited by hypoxia, examples of which have been provided in Table 2. These include classical voltage-gated K^+ channels

(K_V), ether-a-go-go-related K^+ channel (HERG), Ca^{2+} -activated K^+ channel (BK or maxi-K), inwardly rectifying K^+ channels (K_{ir}) and two-pore domain K^+ (K_{2P}) channels. The enigmatic nature of the field of oxygen sensing is in part due to differences in the density and activity of K^+ channel activity which exist between species [80, 81]. Furthermore, developmental change s in oxygen-sensitive K^+ channel expression have been observed, which may contribute to postnatal maturation of oxygen sensitivity [82]. This variation has led researchers to question the innate oxygen sensing abilities of K^+ channels [49, 83, 84]. Conflicting reports of channel function within nearly identical cellular models has made it difficult to agree on a universal mechanism of oxygen sensing (see [83] for a review).

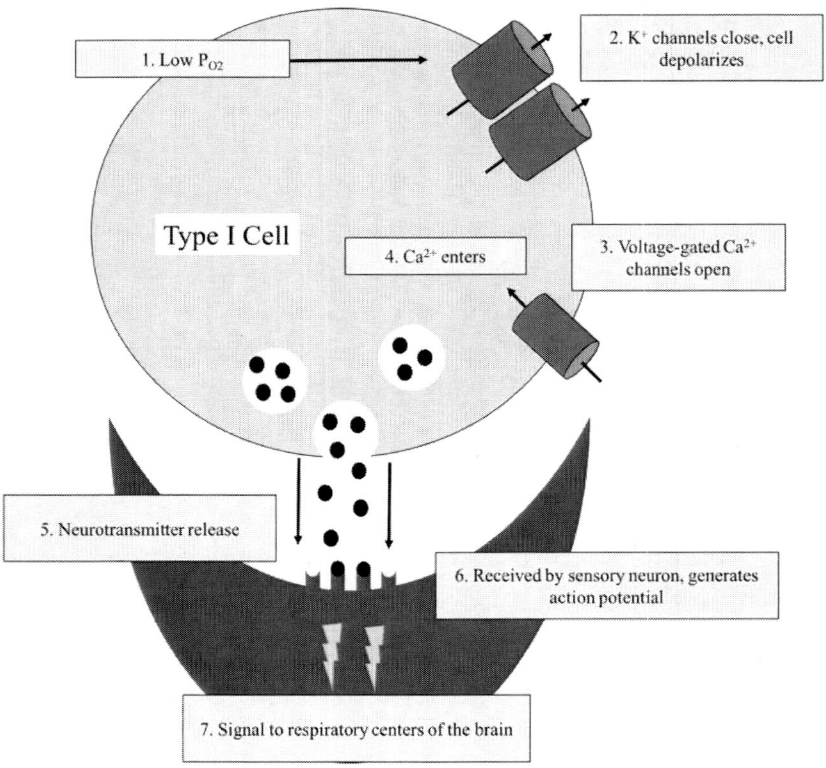

Figure 1. Mechanism of carotid body transduction.

Table 2. Putative oxygen -sensitive K^+ channels in carotid body type I cells

K+ channel identify	Species	References
Inactivating Kv4.1/4.3 channels	Rabbit	(Sanchez et al., 2002)
HERG-like channels	Rabbit	(Overholt et al., 2000)
TASK-like channels (TASK-1/-3)	Rat	(Buckler, 1997)
MaxiK (BKCa) channels	Rat	(Wyatt & Peers, 1995)
Inactivating Kv4.1/4.3 channels	Mouse	(Pérez-García et al., 2004)
MaxiK(BKCa) channels	Mouse	(OTSUBO et al., 2006)
Kv1.2 channels	Mouse	(OTSUBO et al., 2006)
Charybdotoxin insensitive	Cat	(Chou & Shirahata, 1996)
Background (leak; K_B) channels	Zebrafish	(Jonz et al., 2004; Qin et al., 2010)

Despite the constraints of the membrane hypothesis, two details of chemotransduction are clear. First, seminal studies by López-Barneo et al., (1988) and López-López et al., (1989) established that signalling in type I cells depends on sufficient hypoxic stimulus to directly, or indirectly, inactivate at least one type of ion channel in order to depolarize the membrane and stimulate afferent sensory fibres through Ca^{2+} -dependent neurotransmitter. Second, carotid body discharge requires the release of neurotransmitters and the initiation of action potential s in the primary afferent neuron. Additionally, ion-channel inhibition and associated carotid body discharge is proportional to the severity of hypoxia [85]. A promising avenue of investigation came from Lopez-Lopez and Gonzalez, who determined that inhibition of K^+ channel activity could be fully reversed by supplementing hypoxic medium with CO [86]. The consequent antagonism of sensory activity provided convincing evidence that oxygen sensing likely involves a haem protein associated with the plasma membrane [86].

In fish, a study by Jonz et al., (2004) provided the first compelling evidence of the role for fish NECs in oxygen chemoreception, by demonstrating that NECs isolated from zebrafish gill filaments respond to hypoxia through inhibition of background K^+ channels. In NECs, the primary ion channel type appears to be the background (also known as leak) K^+ (K_B) channels [15, 87], which are similar to the TWIK-related tandem pore domain acid -sensitive K^+ channel (TASK)-type ion channel in the mammalian carotid body [88]. Hypoxic inhibition of K_B channels reduces membrane permeability to K^+, which leads to depolarization of the NEC membrane and initiates a brief rise in intracellular Ca^{2+} following activation of voltage-dependent Ca^{2+} channels. Observations by Zachar and Jonz (2017) suggest native long-lasting (L-type) Ca^{2+} channels are important for control of intracellular Ca^{2+}. These authors recorded Ca^{2+} channel activity using patch clamp and fura-2 Ca^{2+} imaging after administering cadmium and a Ca^{2+}channel blocker (nifedipine). The results of this recently published study demonstrate that L-type channel inhibition blocks synaptic vesicle recycling in response to hypoxia in isolated cells [16].

Gasotransmitter Hypothesis

Gasotransmitters are endogenously produced, biologically active gases capable of freely permeating membranes where they transduce chemical signals and initiate physiological or biochemical responses [89]. In animals, three gasotransmitters—NO, CO, and hydrogen sulphide (H_2S)—have been shown to play a signalling role in an array of physiological functions via autocrine, paracrine, or neuroendocrine routes [89-91], and have since received significant attention for their putative role in modulating oxygen chemoreception [92-95]. Studies involving H_2S that have integrated the use of fish models —in particular, the zebrafish (*Danio rerio*)—have provided considerable evidence for hypoxia -mediated cardiorespiratory effects of H_2S [91, 96, 97]. Previous support from a study by Lopez-Lopez and Gonzalez [86], in which it was recognized that CO could modify K^+ channel activity in hypoxia, led researchers to turn their attention to heme oxygenase-2 (HO-2). Using mass spectrometry -base d analysis of oxygen-sensitive MaxiK^+ (BK_{Ca}) channels in rats, Williams et al., revealed that hypoxic inhibition of K^+ currents may be mediated in part by HO-2. This antioxidant enzyme requires nicotinamide adenine dinucleotide phosphate (NADPH) and oxygen as co-factors to catabolize haem into CO, biliverdin, and iron [98]. Under hypoxic conditions, CO generation through HO-2 haem degradation decreases, leading to a reduction in MaxiK$^+$ channel activity, channel closure, membrane potential depolarization, and finally neurotransmitter release [94]. Unfortunately, studies involving HO-2 deficient mice have been challenged by experiments performed on HO-2$^{-/-}$ mice which develop normally and exhibit no change in carotid body secretion or reduced ventilatory drive [99]. Similarly, studies involving HO-2$^{+/+}$, HO-2$^{-/-}$, and HO-2$^{+/-}$ mice were unable to reproduce blunted glomus cell sensitivity to hypoxia in all cases [100]. Nevertheless, hypoxic stimulation of type I cells through changes in [CO] remains one of the best-characterized mechanisms of gasotransmitter oxygen sensing studied to date.

In keeping with mammalian research, two of the three haem oxygenase isoforms have been documented in fish including the inducible HO-1 as well as the constitutively expressed HO-2 [101, 102]. Indeed, emerging evidence in goldfish (*Carassius auratus*) and zebrafish suggests that CO inhibits ventilatory activity via pathways similar to mammals. Tzaneva and Perry (2014) reported that inhibiting HO-1 in goldfish acclimated to 7°C led to reduced CO expression. Interestingly goldfish acclimated to 25°C did not exhibit reduced CO expression despite HO-1 inhibition, which may reflect temperature specific differences in ventilatory responses related to HO-1 expression [102]. HO-1 protein and mRNA have also been reported in the gills of European sea bass (*Dicentrarchus labrax* L.) [103, 104]. The modulatory effects of the HO-CO system on respiration in other fish species warrants further research.

Similar to CO production, the conversion of L-arginine to L-citrulline by nitric oxide synthase (NOS) produces NO in a reaction limited by oxygen. Basal NO released from nerve endings exerts an inhibitory influence on sensory discharge, likely through retrograde signalling on Ca^{2+} channels. Consistent with the idea that NO mediates efferent inhibition, exogenous NO donors have been shown to decrease intracellular Ca^{2+} concentration and inhibit Ca^{2+} currents in glomus cells [105, 106]. NO-induced inhibition of Ca^{2+} channels may in turn activate K^+ channels and thus contribute to a reduced excitation of K^+ channels and carotid body activity. In contrast, NOS activity is reduced upon exposure to hypoxic stress, leading to a decrease in intracellular NO and a reduced inhibition of Ca^{2+} channels. This leads to an increase in $[Ca^{2+}]_i$, cellular depolarization, and neurotransmitter release [105]. Indeed, a considerable body of evidence links linear decreases in NOS activity and blunted ventilatory drive to proportional, physiologically relevant decreases in oxygen tensions [105, 107, 108].

Two nitric oxide isoforms—nNOS and iNOS—homologous to those found in mammalian tissues have been identified in fish [109, 110]. Reports that NOS are present in gill tissue support a role for this enzyme in regulating the ventilatory response to hypoxia in fish; however, its participation in the hyperventilatory responses of fish subject to hypoxia has yet to be firmly established [13, 111].

In mammalian tissues, hydrogen sulphide (H_2S) synthesis is facilitated by endogenous enzymes and by non-enzymatic pathways (e.g., reduction of thiols and thiol-containing molecules). Studies in animals demonstrate its ability to inhibit COX, the terminal enzyme of oxidative phosphorylation in the inner mitochondrial membrane [112, 113]. H_2S has been shown to regulate many diverse physiological and pathological processes, including metabolism, catecholamine secretion, vascular tone, epithelial Na^+ transport, and oxygen chemoreception [114-118]. Its ability to act as a respiratory stimulant has long been appreciated but largely undefined. The discovery that cystathionine γ-lyase (CSE) and cystathionine β-synthase (CBS)—which confer the ability to generate H_2S—are expressed in type I cells suggests that H_2S contributes to activation of the carotid body during reduced oxygen availability [119-121]. Studies have previously demonstrated endogenous H_2S production by CSE and CBS increases 2.5-fold upon onset of hypoxic exposure [117]. Furthermore, genetic deletion and pharmacological inhibition of CSE and CBS in mice produced noticeable reductions in H_2S production. H_2S generation was 90% lower in $CSE^{-/-}$ mice, and this difference corresponded to a similar decrease in both respiratory rate and tidal volume in normoxia as well as hypoxia [117].

Olson et al., (2008) were the first to present evidence showing that H_2S contributes to hypoxic bradycardia and hyperventilation by injecting NaHS into the inspired water. Bilateral extirpation of the externally oriented NECs in the first gill arches of rainbow trout attenuated the hypoxic cardiac response, suggesting that H_2S acts on branchial oxygen chemoreceptors [122]. Additional evidence came from Porteus et al., (2014) who observed that both CBS and CSE were expressed in the NECs of zebrafish and that exogenous H_2S induced hyperventilation in a dose-dependent manner correlated with membrane depolarization and an increase in intracellular Ca^{2+} in NECs. Furthermore, these authors also showed that adult zebrafish display reduced H_2S production and a blunted ventilatory response following pharmacological inhibition of CSE and CBS [97]. In larvae, CBS and CSE antagonism completely abolish the ventilatory response. Similar studies which supplemented ventilated birds with H_2S reported an increase in the discharge frequency of intrapulmonary CO_2 receptors as well as an increase in breathing amplitude [123]. Further research into the role of H_2S in regulating hypoxic responses in avian models is lacking.

A consensus regarding the nature of the signalling pathways that link H_2S to changes in ion channel activity has yet to be reached. Two recent studies by Li et al., (2010) and Telezhkin et al., (2010) provide unequivocal evidence for the role of H_2S in hypoxic signalling by demonstrating that application of donor H_2S rapidly inhibits MaxiK$^+$ channels and elevates cytosolic Ca^{2+} in mice glomus cells [118, 124]. These findings also support the notion that H_2S excitation of carotid body chemoreceptors may involve covalent modification of redox sensitive cysteine residues in L-type Ca^{2+} channels via *S*-sulfhydration [125]. The idea that H_2S can directly influence the activity of K^+ channels was recently brought into question by Buckler et al., (2012), who noted the similarity between the effects of H_2S and other inhibitors of mitochondrial COX [126]. H_2S is a well-known inhibitor of oxidative phosphorylation and this action has previously been attributed to the activation of ATP - sensitive K^+ channels through a decline in cellular ATP levels [113, 127]. Buckler's lab revealed a parallel between the effects of H_2S on $[Ca^{2+}]_i$ and its effect on another index of mitochondrial function: $[Mg^{2+}]_I$ [126]. They propose two potential pathways via which H_2S acts on mitochondrial energy metabolism, rather than directly upon oxygen sensitive K^+ channels: the first being direct modulation of ATP-sensitive TASK channels by changes in MgATP levels, and the second being indirect modulation via changes in AMP/ATP levels and AMP kinase, which will be discussed below.

Protein Kinase G-Regulated Production of H_2S

The gasotransmitter hypothesis of carotid body transmission via ion channel modification is an idea which has garnered support for many years. However, until recently, little was known about the sensitivity of heme -oxygenase and the mechanism by which CO influences K^+ channels. Furthermore, evidence that multiple gas messengers were capable of affecting ventilatory drive cast doubt on their role in oxygen sensing. In a recent study, however, Yuan et al., (2015) proposed a unified mechanism for how the three gasotransmitters work in tandem to regulate breathing (Figure 3) [128]. These authors showed that two conserved cysteine residues (Cys^{265} and Cys^{282}) confer HO-2 with low oxygen affinity, which allows it to translate changes in oxygen availability into changes in CO generation. The authors also established that CSE-dependent H_2S generation is not intrinsically oxygen sensitive by transfecting human embryonic kidney -293 (HEK-293) cells lacking functional HO-2 with mouse CSE, thereby confirming the requirement for a secondary messaging system. To reconcile these findings, Yuan et al., (2015) hypothesized CO is actually a physiological inhibitor of CSE through soluble guanylyl cyclase (sGC)/cyclic guanosine monophosphate (cGMP) signalling. Hypoxia elicits HO-2 dependent reductions in CO, and subsequently decreases sGC/cGMP concentrations, thereby preventing CO-inhibition of CSE and H_2S generation. To confirm the activation of CSE by CO and secondary messengers, the authors subjected cell line s co-expressing HO-2 and CSE to inhibitors of both sGC and cGMP. Small interfering RNA -mediated silencing of sGC and cGMP prevented HO-2 inhibition of H_2S generation under normoxia.

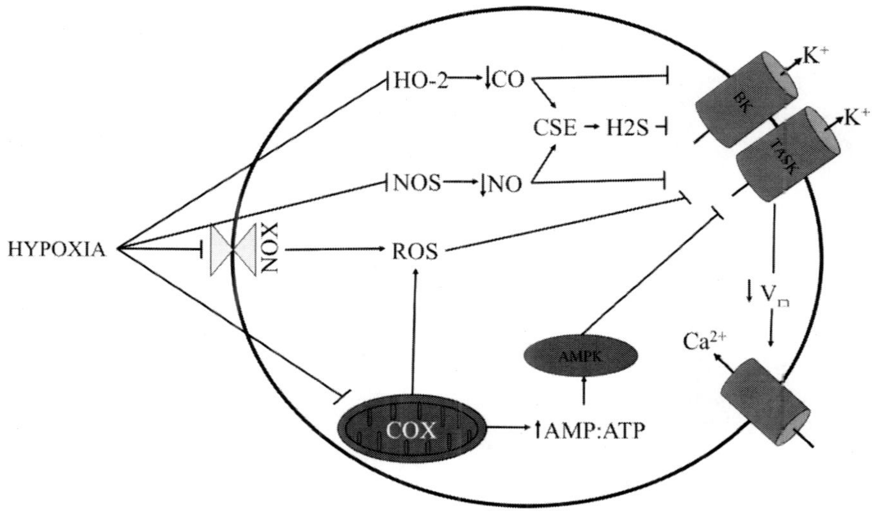

Figure 2. Putative mechanisms of hypoxic transduction in the carotid body. Arrow indicates activation or upregulation, hammer indicates inhibition or downregulation.

In other biological systems, activation of sGC/cGMP signalling pathways stimulates cGMP-dependent protein kinase G (PKG) phosphorylation of serine and threonine residues in target proteins [129]. To determine if PKG-dependent phosphorylation of CSE is necessary for its inactivation in hypoxia, these authors treated CSE-expressing cells in hypoxia as well as with a CO donor and measured changes in phosphoserine levels. While exogenous CO increased serine phosphorylation of CSE, hypoxia reduced phosphoserine dramatically. Furthermore, inhibition of either sGC or PKG prevented CSE phosphorylation in CSE/HO-2 cells under normoxia. Through sequence analysis, the group discovered a highly conserved putative PKG recognition sequence containing a conserved serine residue (Ser^{377}). Mutation of Ser^{377} with alanine completely abolished PKG-mediated phosphorylation of CSE.

Taken together, CO generation by oxygen -sensitive HO-2 inhibits CSE activity through sGC/cGMP stimulation of PKG. Preventing any step in the pathway leads to an increase in H_2S generation and subsequent increase in sensory nerve activity similar to that observed in hypoxia. Interestingly, mice deficient in HO-2 still exhibited hypoxia-induced changes in H_2S concentrations and an increase in carotid body discharge suggesting the existence of an alternative regulatory mechanism that does not involve HO-2. Indeed, Yuan et al., discovered that NOS is upregulated in HO-2 knockout mice. NO inhibits carotid body sensory nerve activity through the same sGC/cGMP and PKG pathways to regulate H_2S and thus acts as a substitute for CO when HO-2 is absent. Overall these findings elegantly highlight the importance of built-in redundancy within oxygen-sensing systems.

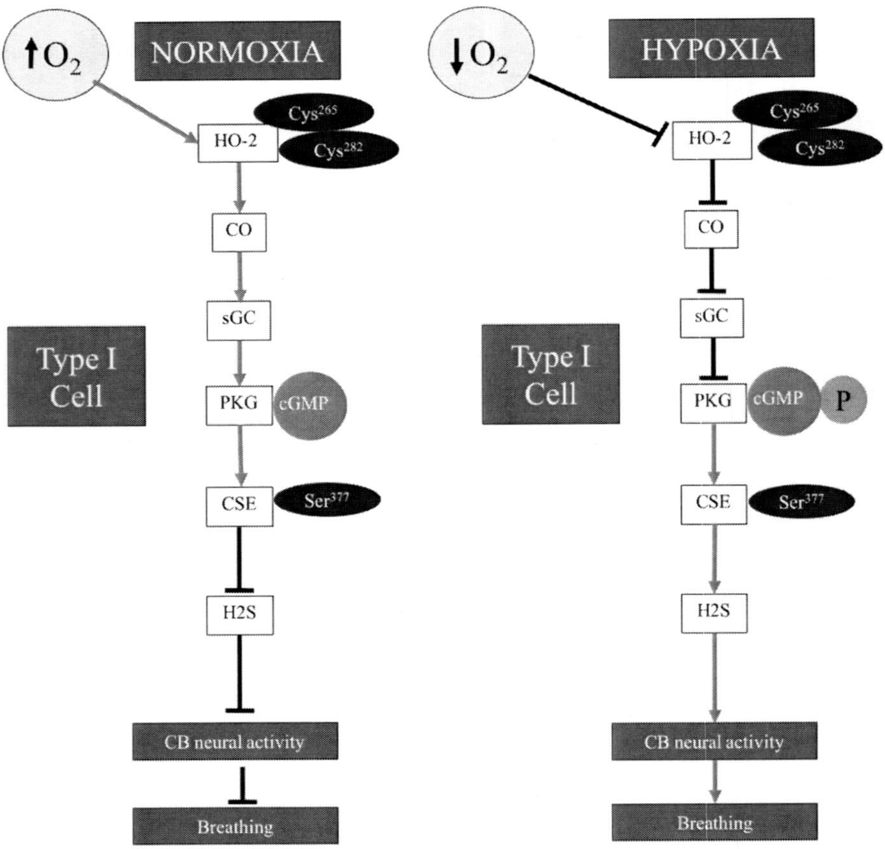

Figure 3. Gasotransmitter hypothesis proposed by Yuan et al., 2015.

Globins as Oxygen Meters

It is well established that red blood cells facilitate the transport of sufficient oxygen between respiratory surfaces (lungs, gills), and respiring tissues via haemoglobin and related allosteric interactions between ligand (O_2, CO_2, H^+) binding sites. The primary function of globins as carriers of oxygen has led some to postulate a possible contribution to oxygen homeostasis by coupling changes in the degree of deoxygenation to the release of vasodilatory compounds such as ATP or NO [130, 131]. Biochemical evidence that haemoglobin and myoglobin exhibit enzymatic behaviour as a nitrite reductase, suggests they may play a role in oxygen sensing and hypoxic vasodilation [132, 133]. NO plays a fundamental role in coordinating normal vasomotor tone among other physiological processes [134, 135]. Previous studies support a model whereby red blood cells mediate vasoactivity through the transformation of nitrite to NO by a deoxyhemoglobin-mediated reduction reaction [131, 136]. Evidence that haemoglobin-mediated NO production participates in the hypoxic vasodilatory response comes from Crawford et al., (2006), who demonstrated that deoxygenating RBCs and nitrite produce NO and stimulate vasodilation, cGMP formation, and the inhibition of mitochondrial respiration [137]. This pathway has been the subject of an excellent recent review: [138].

NADPH Oxidase

Biological complexes that rely on oxygen in order to function may be the first step in coupling reductions in PO_2 with carotid body transduction. A number of research groups have proposed that NADPH oxidases (NOX), a family of haem-containing, membrane-bound enzyme complexes, regulate ion channel conductivity through its relationship with molecular oxygen, although compelling evidence supporting their relationship with carotid body transmission has yet to be presented. According to the

NADPH model of oxygen sensing, NOX converts changes in oxygen into a redox signal through generation of superoxide (O_2^-) [139], which, after dismutation to hydrogen peroxide (H_2O_2), alters the cytosolic redox status. ROS modify target protein function via oxidation of reactive thiol groups, and therefore are well-equipped to regulate oxygen signalling pathways [140, 141]. Presumably, a decrease in oxygen availability would lead to a decrease in ROS generation and a shift in the cytosol to a more reduced state. This system would alter K^+ channel conductance without direct inhibition of the channel itself. Ultimately this would cause a shift in membrane potential, causing Ca^{2+} influx into chemoreceptor cells and the release of neurotransmitters [142].

The expression of NOX in cells identified as oxygen responsive—such as NEBs of the lung [52], type I cells of the carotid body [143], and pulmonary vascular myocytes [144]—previously supported their involvement in mediating hypoxic responses [55]. Multiple isoforms of NOX have been offered as potential candidates for oxygen sensing, and these include a multi-subunit assembly of p22phox, gp91phox, p47phox, p47phox, p40phox, p67phox, and Rac1, 2 subunits. Mice with selective gp91phox subunit deletion exhibit reduced oxygen sensitivity of NEB K^+ channel conductivity and impaired ventilatory control but retain normal PASMC and carotid body function [145-147]. Similarly, activation of NOX causes hypoxia -like K^+ channel closure in NEB H146 cell line s. In contrast, NOX inhibition fails to suppress the hypoxic response of the carotid body, nor does it affect the hypoxia-induced release of catecholamine commonly observed in this system. Further refuting the role of NOX as an oxygen sensor, selective deletion of Nox subunit p47phox mice resulted in enhanced sensory activity within the carotid body [148, 149]. Available literature is restricted to a few species of mammals ; however, what data there is points toward potential cell type specific involvement of NOX in detecting oxygen.

O_2-Sensitive Ca^{2+} Channels

The central involvement of voltage-gated Ca^{2+} channels in neural pathways and the fast adaptation of cardiorespiratory systems make these proteins enticing potential candidates for direct-activation by changes in blood oxygenation status. Indeed recent studies have provided evidence for oxygen -sensitive native L-type Ca^{2+} channels that are rapidly and reversibly inhibited by reductions in arterial oxygen content [150, 151]. The contribution of L-type channels to increased cytosolic Ca^{2+} during hypoxia has been extensively detailed in PASMCs and may account for arterial dilation during exposure to hypoxia [151-154]. While similar L-type channels have been documented within the carotid body in both rabbits and rats [155-157], follow up literature investigating their role in chemoreception has yet to be conducted.

Na^+ Channels

Although K^+ channels are classically considered to be the primary regulators of type I cell chemotransduction, it is possible that other types of inward ion currents may be involved in hypoxia -induced membrane depolarization. Reductions in oxygen availability could increase background Na^+ currents, although to our knowledge this process has not been definitively characterized. Carpenter and Peers (2001) suggest K^+ channel closure alone cannot adequately explain membrane depolarization in response to hypoxia, and propose a separate depolarizing factor must exist in chemoreceptive cells. These authors postulate the resting (background) Na^+ influx is necessary for depolarization. Indeed, they recorded a standing Na^+ current in type I cells isolated from the rat carotid body [158]. Further research is warranted to flush out our understanding of the putative role for Na^+ channels in oxygen sensing and chemotransduction.

MITOCHONDRIA AS OXYGEN SENSORS

Mitochondria have long been hypothesized to be mediators of hypoxic chemoreception [159]. Classic studies involving genetic and/or pharmacological interventions targeting the mitochondrial electron transport system (ETS) and all uncouplers of oxidative phosphorylation mimic hypoxia through stimulation of peripheral chemoreceptors to induce neurosecretion [160-163]. Conceptually, mitochondrial regulation of oxygen signalling pathways is intuitive: activity of the ETS ceases in the absence of oxygen, leading to reduced ATP generation and changes in cellular ion gradients. Theoretically, these changes allow mitochondria to translate low oxygen stress into signalling intermediates, which mediate cytoprotective responses such as synaptic function, gene transcription, and/or entry into cell death pathways. However, evidence challenges their ability to detect physiologically relevant changes in oxygen tension by demonstrating COX activity does not become oxygen limited until PO_2 reaches near-anoxic levels («0.1% O_2), a level well below that needed to initiate acute and/or chronic physiological responses [164-166]. This would suggest mitochondria are ill-equipped to serve as sensors of moderate hypoxia. However, low affinity mitochondria have been detected by a variety of research groups and thus it remains possible that the oxygen sensing capacities of mitochondria are limited to certain cell types [167, 168].

Mitochondrial Hypothesis

To reconcile the kinetic behaviour of mitochondrial cytochromes in shaping adaptive responses, Wilson et al., investigated the relationship between low oxygen stress and redox changes in COX [165, 169]. COX is the terminal complex located on the inner membrane of the mitochondrial ETS and contains two heme (*a* and *a3)* and two copper centers (Cu_A and Cu_B). Combined, COX a_3 and Cu_B constitute the binding site for oxygen, which, upon binding, catalyzes the oxidation of COX and the reduction of oxygen to water. When oxygen is limiting (<1 mmHg), Wilson et al., observed an equivalent increase in the reduction state of COX [165, 169]. While these studies failed to characterize a cellular response, their results broke ground for a possible 2nd messenger system coupled to redox changes in the ETS. A decade later, Chandel et al., discovered that COX exposed to 1-2 h of hypoxia developed a change in its catalytic activity by decreasing its maximal turnover rate. This change in oxidase behaviour is coupled to decreases in mitochondrial respiration caused by a reduction in cellular metabolic demand rather than a dysfunctional ETS [170, 171]. The team proposed a reversible depression of the maximum rate at which an enzyme can catalyze a reaction; V_{max}—of COX may convert the mitochondrial ETS into a low PO_2 affinity system to confer sensitivity over a physiologically significant range, although it is still unclear how this signal is transmitted throughout the cell.

Further support for the mitochondria hypothesis comes from Biscoe et al., who demonstrated that hypoxia causes mitochondrial depolarization in glomus cells at much higher P_{O2} values that in mitochondria from other tissues [172]. Similarly, previously measured K_m values for COX *in vivo* are 20 to 30 times higher than that of the isolated enzyme what may suggest cooperation between different sensors [173]. According to these observations, glomus cell mitochondria may be sensitive to small changes in oxygen compared to other neuronal cells. It has been suggested that NO is an endogenous mediator which interacts with COX to confer low oxygen affinity through competitive binding with oxygen [174, 175]. However, this lends more support to the gasotransmitter hypothesis of oxygen sensing than the mitochondrial hypothesis.

Mitochondrial ROS

Oxygen-dependent changes in the concentration of ROS formed by the incomplete transfer of electrons through mitochondrial complexes I and III constitute an alternative to NOX signalling that would still account for changes in redox status [176-178]. Above a certain threshold, ROS can overwhelm antioxidant defences and cause oxidative stress. However, in a healthy cell, non-harmful levels of ROS play important physiological signalling roles and can mediate cytoprotective pathways against hypoxic and ischemic stress [139, 179-183]. There is a continuing debate as to whether or not ROS production increases or decreases in hypoxia, or whether or not there is a specific ROS species which elicits hypoxic responses. It was traditionally accepted that the concentration of ROS changes in direct relation to cellular oxygen availability. For example, regional reductions in ROS corresponding to low oxygen have been documented in certain anoxia -tolerant species including the brain of freshwater painted turtles (*Chrysemys picta*) and the Common Slider (*Trachemys scripta*) [184, 185], embryos of killifish *(Austrofundulus limnaeus)* [186], heart of epaulette shark (*Hemiscyllum ocellatum*) [187], and hypoxia- adapted Drosophila lines [188]. However, several of these examples involve species or tissues exposed to acute anoxia; since ROS generation requires the presence of oxygen, a decrease in ROS generation in anoxia is perhaps to be expected. In contrast, Chandel and colleagues have consistently reported an increase in ROS production during acute hypoxia [176, 178, 189, 190]. In support of this, recent evidence suggests that increases in transcription and activation of the pro-oxidant enzyme NADPH oxidase-2 contributes to an increase in ROS levels in intermittent hypoxia [191-193]. The effect of ROS on transcription will be discussed in more detail below.

The impact of ROS on K^+ channel function has been described in a number of cell types, particularly in PASMCs [194]; however, the specific molecular target which couples their production with acute hypoxic changes in K^+ channel gating and increases in systemic survival is not well established. A likely target for this system is post-translational protein modifications. Indeed, there is considerable evidence that H_2O_2 is capable of oxidizing thiol groups (—SH) on cysteine residues. The resulting sulphenic acid (—SOH) can react with glutathione or with amides, to form ether disulphide bonds (—SS—) or sulfenyl amide (—SN—) respectively [195, 196]. These alterations have the capacity to change target-protein function, which may include inhibiting K^+ channels within the carotid body or PASMCs [141].

Experiments conducted by Bernardini et al., (2015) propose carotid body type I cells possess several systems capable of generating ROS. They confirmed one to be NOX, which exhibit impaired ROS production in response to hypoxia. These authors suggested that the other is likely to be either mitochondria or endoplasmic reticulum oxidoreductin-1, which responds to hypoxia with an increase in ROS production [197, 198]. The two putative systems allegedly can be manipulated under different conditions, but likely interact with each other to keep tight regulation over carotid body sensory activity. Additionally, the authors reported that only NADPH-mediated reductions in ROS stimulated membrane depolarization, even in the presence of a predominant hypoxic ROS increase, supporting the hypothesis that ROS transfers oxygen sensitivity to K^+ channels to regulate ventilatory drive. In support of this, TASK1 oxygen sensitivity was augmented by NOX4 overexpression and was inhibited through NOX4 antagonism [197, 198].

Much of the disagreement in the field over the role and source of ROS in oxygen sensing may be due to the technically challenging nature of ROS visualization and quantification, which often generates contradictory data. Regional changes in ROS, which may selectively regulate hypoxic responses, are often obscured by limited technology and the very rapid degradation of ROS within the cell. Putative contributions of ROS to altered carotid body function remain to be investigated.

Metabolic Hypothesis/Adenylate Charge Hypothesis

Adenosine monophosphate-activated protein kinase (AMPK) is a well-known metabolic stress sensor and regulator of protein synthesis expressed in nearly all cells; a recent study proposed that increased mitochondrial

ROS production excites the carotid body through activation of AMPK [199-201]. AMPK activation by ROS would subsequently inhibit oxygen sensitive K^+ channels [201]. This notion has recently been challenged, however, and it has since been suggested that AMPK activation in the carotid body is coupled to mitochondrial respiration through alterations in the cellular AMP:ATP ratios resulting from inhibition of oxidative phosphorylation during periods of hypoxia. Specifically, during periods of hypoxic stress, and consequently reduced ATP, the AMP:ATP ratio is increased which in turn activates AMPK. AMPK appears to cycle through phosphorylation states, ultimately returning to a dephosphorylated, inactive form in normal conditions [202]. The role of AMPK in carotid body sensing was observed through pharmacological AMPK agonism, which triggers Ca^{2+} influx in isolated rat type I cells and increases afferent sensory discharge—mirroring hypoxia [203, 204]. Further experiments revealed that AMPK directly phosphorylates both recombinant BK_{Ca} and leak K^+ currents, suggesting a possible mechanism for the modulation of oxygen-sensitive ion channel activity [205, 206].

AMPK becomes an even more attractive candidate when one considers its role in the coordination of metabolic responses to hypoxia. When AMPK is activated it regulates a variety of cellular metabolic systems. For example, it stimulates fatty acid oxidation and glucose uptake, while inhibiting protein, fatty acid, glycogen and cholesterol synthesis. AMPK acts to inhibit anabolic energy consuming pathways in the cell including protein synthesis via the phosphorylation of elongation factor 2, while also stimulating energy production via catabolic pathways [207, 208]. For example, Jibb and Richards presented evidence that AMPK plays a role in hypoxia tolerance in facultative anerobes by demonstrating that severe hypoxic exposure for 1-12 hours increased AMPK activity and decreases protein synthesis and elongation factor 2 expression in the liver of goldfish [209]. Taken together, these observations provide a possible integrated mechanism for both the mitochondrial and membrane hypotheses for carotid body oxygen sensing.

HYPOXIA-INDUCIBLE FACTORS

Many homeostatic responses to extended bouts of hypoxia involve changes in gene activity and protein expression achieved through a coordinated relationship between several transcription factors, transcriptional coactivators, and DNA -dependent RNA polymerases. The conditions underlying gene regulation maintain that transcription factors must: a) be capable of regulation themselves, and b) possess recognition site specificity within gene-target binding domains. Hypoxia-inducible factors (HIFs) are a highly conserved group of transcriptional activators first identified in a pioneer study led by Semenza and Wang in 1992. These authors discovered a 50-nucleotide enhancer in the 3' flanking region of the human erythropoietin gene which increased transcriptional induction of the gene sevenfold [210]. The discovery that transcription factors binding to this sequence could be induced by hypoxia led to an avalanche of new insights into oxygen -regulated gene expression [211, 212]. In meeting the aforementioned prerequisites for transcription, HIF activity is inducible in an oxygen-dependent, posttranslational manner and binds to DNA sites represented by a core consensus sequence (5'-RCGTG-3'), located within a flanking hypoxia response element (HRE) [213].

Hypoxic HIF stabilization and target gene transcription has received considerable research attention and the mode of action of this system is very well known. Furthermore, the biological importance of HIFs is reinforced by their central role in developmental, physiological, and pathophysiological responses to hypoxia. Analysis of HIF-1α null mice (*Hif*$^{-/-}$) reveals embryonic lethality (between E8.5 and E10.5) with apoptosis of the neural mesenchyme and aberrant vasculature compared with wild-type mice [214, 215]. Mice possessing only one mutant allele (*Hif*$^{+/-}$) develop normally, but display impaired physiological responses to chronic hypoxia [216]. In addition to development, in vivo studies have determined HIF induction is associated with specific aspects of pathological conditions involving tissue hypoxia, including cardiovascular disease, inflammation, anemia, and cancer.

Structure, Stability, and O_2 Sensing

The transcriptionally active form of HIF is a dimeric protein complex consisting of one of three oxygen -labile alpha subunits (HIF1α, HIF2α, or HIF3α) and a beta subunit called aryl hydrocarbon receptor nuclear translocator (ARNT) (Figure 4). Both HIFα and HIFβ belong to the basic helix-loop-helix (bHLH)/Per-ARNT-Sim (Pas) family of transcription factors. The PAS domain contains approximately 300 amino acid s and operates in conjunction with the primary HLH interface to mediate protein-protein interactions and α/β dimerization [212, 217]. Molecular comparisons of HIF1α and HIF2α reveal nearly identical biochemical properties (i.e., modular protein structure, hypoxic protein stabilization, dimerization with ARNT, and overlapping DNA recognition sequences), however HIF-2α and HIF3α exhibit tissue -specific expression patterns restricted to distinct cell types. Studies of these different isoforms have shown they control separate, non-redundant biological functions during embryonic development [218]. Furthermore, evidence has recently been provided that mutual antagonism between HIF1α and HIF2α may regulate oxygen sensing and cardiorespiratory homeostasis through a delicate ROS balance [219].

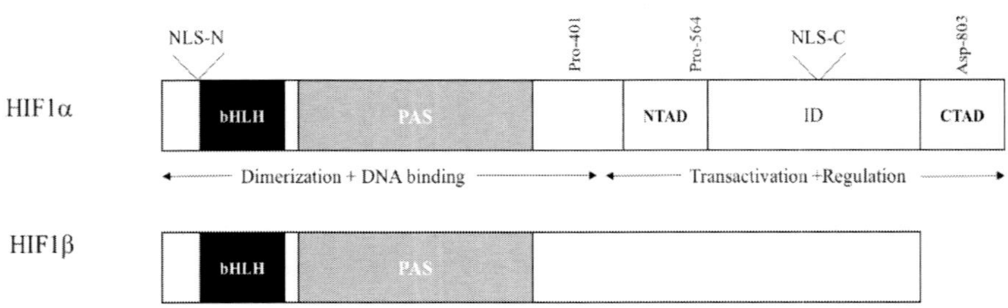

Figure 4. Structure of (human) hypoxia -inducible factor -1 modified from (Semenza, 2000).

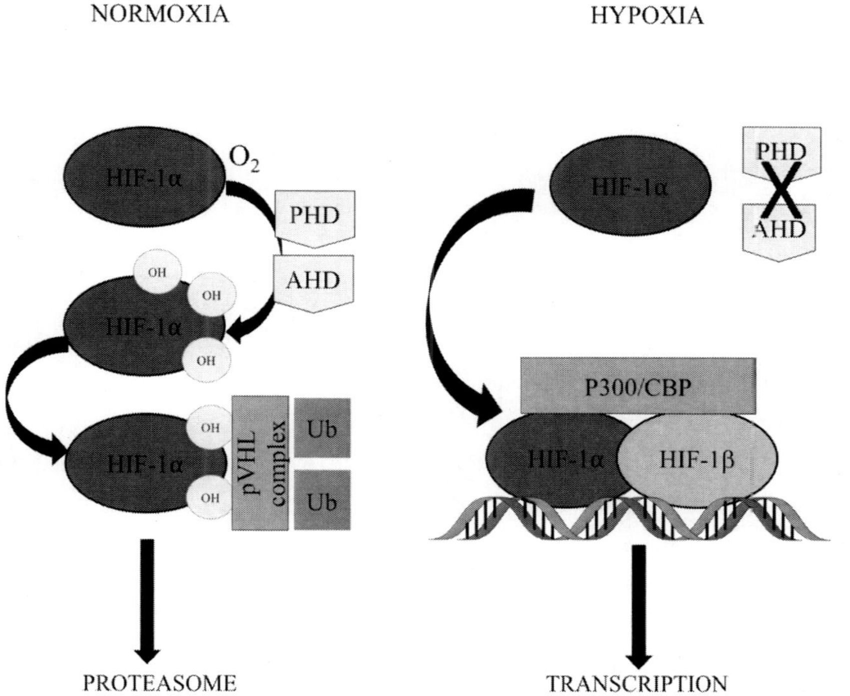

Figure 5. HIF stabilization in hypoxia.

Under well-ox ygenated conditions, HIFα is rapidly hydroxylated on at least one of two conserved proline residue within the oxygen -dependent degradation domain (Pro-402 and/or Pro-564) by members of the EGL-9/prolyl hydroxylase (PHD) superfamily of Fe(II)- and 2-oxoglutarate-dependent dioxygenases [220]. Hydroxyproline substitutions within the ODD render HIFα subunits vulnerable to proteasomal degradation under normoxic conditions. Three novel HIF prolyl hydroxylase enzymes (PHDs 1,2,3) have been identified, of which PHD-2 appears to be the most important in regulating α-subunits [220]. Once hydroxylated, HIFα is captured by the β-domain of the von Hippel-Lindau tumour suppressor protein (pVHL), the substrate recognition component of the Elongin BC/Cul2/pVHL (E3) ubiquitin -ligase complex [221-223]. As a result, HIFα is polyubiquitylated and subjected to proteasomal degradation by the 26S proteasome.

VHL disease is a hereditary cancer syndrome first discovered in 1894, the hallmark of which is a high degree of vascularization in the brain and spinal cord leading to hemangioblastomas. The potential connection between dysfunctional pVHL and the constitutive expression of a large number of hypoxia inducible genes such as vascular endothelin growth factor (VEGF) led to the discovery that VHL was indeed necessary for proteasomal degradation of HIF-1a [224]. Studies have supported its participation in the negative regulation of HIFα by demonstrating biallelic or genetic inactivation ($VHL^{-/-}$) as well as pharmacological antagonism of VHL stabilizes HIFα in normoxia [221, 224].

Prolyl hydroxylases require Fe^{2+}, dioxygen, and 2-oxoglutarate to yield the proline hydroxyl groups. During hydroxylation, one oxygen atom is used in the oxidative decarboxylation of 2-oxogultarate yielding succinate and carbon dioxide, and the other is incorporated directly into the hydroxyl group bound to HIF1α [225, 226]. Therefore, under oxygen-deprived conditions PHD catalytic activity is interrupted, allowing HIFα to escape recognition by the pVHL ubiquitin -ligase complex [223]. When stabilized, HIFα is translocated to the nucleus, where it associates with nuclear ARNT. The HIF dimer then recruits the general transcriptional activator CREB binding protein (CBP)/p300 which interact with the carboxy-terminal transactivation domain, and the completed complex binds to hypoxia response elements (HREs) present in the promoter region of hypoxia-inducible genes that offer protection against oxygen limitation (Figure 4).

Notably, asparaginyl hydroxylase—also known as factor-inhibiting HIF (FIH)—further contributes to oxygen -dependent regulation of HIF [223, 227]. At sufficient oxygen tensions, an asparagine residue (ASP-803 in humans) is hydroxylated, preventing the interaction of HIF with CBP /p300 and reducing binding activity with the HRE [228]. FIH1 remains active at lower oxygen concentrations than prolyl hydroxylase suggesting these two enzymes may fine-tune HIF functional activity within distinct tissue -types or target genes at different oxygen-dependence profiles [228]. Additionally, the different affinities of PHDs and AHDs for oxygen may explain differences in gene function and adaptive responses in different species groups. Together, the depth of hypoxia required to stimulate a given HIF-responsive gene depends on isoform activity and/or cross-talk with other signaling pathway s.

By virtue of their compulsory use of oxygen, PHDs and AHDs function as conserved physiological oxygen meters. Epstein et al., (2001) demonstrated that graded decreases in oxygen yielded parallel and progressive increases in the amount of stabilized HIFα *in vitro* [220]. Similarly, *in vivo* immunohistochemical staining reveals low HIFα levels in rodent tissues subjected to normal physiological conditions and a dramatic increase in staining in response to systemic hypoxia [229]. Stroka et al., (2001) and Wiesener et al., (2003) also demonstrated differential HIFα expression among different organs supporting the model of tissue - specific adaptive protein synthesis previously proposed [229-232]. The exact oxygen threshold required to inhibit PHDs and stabilize HIF varies between individual cell types, although in mammalian cells this threshold is typically lower than that required to initiate membrane depolarization and carotid body activity.

HIF and ROS

The first evidence that mitochondrial ROS are capable of stabilizing HIFα came from studies in the late 1990s involving cells lacking mitochondrial DNA (p^o cells) [233]. p^o cells lacking a functional ETS or wild-type cells treated with an inhibitor of the respiratory chain could not generate mitochondrial ROS under hypoxia [176, 233]. In support of this, genetic knockouts for complex III exhibited increased ROS generation and showed enhanced accumulation of unhydroxylated HIF-1α [191]. To assess whether hypoxia-induced decreases in oxidative phosphorylation, or increases in mitochondrial ROS are needed to stabilize HIF-1α, Bell et al., (2007) used cells deficient in cytochrome b which were unable to perform mitochondrial respiration, but still produced mitochondrial ROS [234]. These cytochrome b-null cells exhibited a normal HIF-response to hypoxia which could be inhibited by pharmacological ROS antagonists.

The precise mechanism of HIF-1α stabilization during low oxygen stress remains poorly understood but is thought to involve mitochondria - derived ROS since free radical scavengers abolish the hypoxic HIF response [233]. A leading model to explain HIF-1 stabilization by mitochondria posits that during hypoxia, increases in mitochondria-generated ROS prevent HIF-1α protein degradation and thereby communicate a change in oxygen availability to the HIF machinery. This subsequently activates the myriad of cellular responses to hypoxia mediated by HIFs [178, 233].

Adaptive Hypoxic Responses and Their Regulation by HIF

Depending on the length of exposure, vertebrates faced with a regional or systemic hypoxic challenge may exhibit several short and/or long term physiological responses that help to ameliorate the effects of low oxygen availability [11]. Short term responses typically involve reflex modifications to ventilation, cardiac output, metabolism, and oxygen carrying capacity, with the immediate goal of increasing oxygen delivery to tissues. Short term responses may also include a temporary switch from aerobic to anaerobic metabolism, as well as a global suppression of nonessential protein synthesis in order to reduce metabolic demand and restore homeostasis. During chronic hypoxic exposures, vertebrates typically experience further increases in whole body ventilation, prolonged and deepened metabolic depression, and increased expression of hypoxic survival genes mediated by HIFs. These responses to hypoxia are diverse and vary between species, throughout development, and in wake versus sleep states, and are plastic depending on the degree and duration of the hypoxic exposure. In general, however, these adaptations mitigate the impact of low oxygen stress by increasing oxygen uptake into the body, maximizing delivery to tissues, and/or reducing metabolic demand [200, 235]. HIF plays a central role in regulating physiological adaptations and survival responses to acute and prolonged hypoxia, for example, genes activated by HIF transcription mediate angiogenesis, anaerobic glycolysis, cellular proliferation, and neural stem cell growth. The impact of HIF on these systems is mediated by a wide variety of gene targets; indeed, it is estimated that the complete HIF transcriptome likely contains thousands of genes [236].

Increasing Delivery: The Hypoxic Ventilatory Response

A common physiological response to hypoxia is to increase ventilation in an effort to increase the amount of oxygen reaching the tissues. This hypoxic ventilatory response (HVR) is a common phenomenon among vertebrates and is characterized by dynamic changes in respiratory motor output—the amplitude and direction of which depends on the pattern, duration, and intensity of the hypoxic stimulus [237, 238]. The acute HVR involves an immediate increase in respiratory activity within one to two breaths in hypoxia, and commonly involves increases in both tidal volume and respiratory rate (but see also [239, 240]). This reflex increase in ventilation is mediated by carotid body afferents which travel along the carotid sinus nerve, synapse in the nucleus tractus solitarii of the CNS, and relay carotid body-detected changes in oxygen to the respiratory control centres in the brain to increase ventilation [241-243].

With prolonged hypoxia, most vertebrates exhibit ventilatory plasticity termed ventilatory acclimatization to hypoxia (VAH), which is a time-dependent increase in respiratory activity [237]. The progressive changes in respiratory phenotype during VAH are the result of multiple cellular and synaptic mechanisms that facilitate afferent and efferent communication within the respiratory control system [244, 245]. The respiratory plasticity of VAH is characterized by relatively slow onset and by its persistence despite a return to normoxia, although the exact time course of VAH is species dependent. Changes in carotid body sensitivity play a central role in VAH. Following several days of hypobaric hypoxia, carotid body responses to acute hypoxia are enhanced in cats [246, 247] and rats [248].

In goats, increases in chemoreceptor afferent activity are observed after only 6 hours of hypoxia. Furthermore, ablation or partial dysfunction of the carotid bodies prevents or attenuates VAH in experimental animals and humans [244, 249-251]. Prolonged exposure to several weeks of hypoxia may eventually desensitize the carotid body response to acute hypoxia [252, 253]. In addition to enhanced carotid body sensitivity, VAH involves an increase in the central gain of the respiratory brainstem, where synaptic connections between respiratory nerve tract relays are enhanced. Specifically phosphorylation of glutamatergic receptors in the NTS mediate this central gain in VAH [254-256], and this is likely a commonly conserved mechanism underlying VAH in most mammals (but see also [240, 257]), and represents a form of functional plasticity in the respiratory system following chronic sustained hypoxia.

As introduced above, the mechanisms underlying hypoxic respiratory plasticity involve changes in structural remodeling of the carotid body, K^+ and Ca^{2+} channel function, and altered CNS processing of arterial chemoreceptor afferent input [258, 259]. VAH occurs in response to sustained hypoxia lasting hours to months, and as such it is likely these changes are coordinated by transcriptional changes mediated by HIF1α. Indeed, HIF1α expression increases in the carotid body and CNS respiratory centers of mice after as little as 1 hour of hypoxic exposure [216, 260], mice partially deficient in HIF1α (*Hif*$^{+/-}$) and acclimatized to hypoxia exhibit impaired oxygen sensing by the carotid body, characterized by a reduction in respiratory frequency and depressed neural response [216], and transgenic mice with a CNS-specific knockout of the HIF1α gene but with fully functional carotid body HIF1α do not show an increase in respiratory sensitivity after chronic hypoxia [261, 262]. More recently Peng et al., (2011) demonstrated mice haploinsufficient for HIF2α and subjected to hypoxia experienced an augmented carotid body response, irregular breathing, hypertension, and elevated plasma norepinephrine levels [263]. In addition, HIF2$\alpha^{+/-}$ mice treated with an antioxidant resumed normal carotid body activity and autonomic function, suggesting hypoxia exerts drastically different effects on HIF1 and HIF2 but that both play major roles in mediating physiological responses to long-term hypoxia.

Recent findings demonstrate the increase in carotid body sensitivity is potentially attributable to an increase in cellular redox state through HIF-mediated transcriptional upregulation of pro-oxidants (e.g., NOX) and concomitant downregulation of antioxidant enzyme (e.g., superoxide dismutase-2) [193, 264, 265]. It has been suggested that plasticity of the carotid body in response to chronic intermittent hypoxia is mediated by positive feedback interactions between HIF-1α and ROS. As previously discussed, NOX may function to increase ROS generation in hypoxia, which in turn stabilizes HIF1α and downregulates HIF2α [192]. Once activated, HIF1α may induce further increases in ROS by increased expression of NOX genes [264]. Taken together, these findings provide further support for ROS-mediated oxygen sensing and signaling in the carotid body.

Anatomically, the carotid body grows to twice its size in chronic hypoxia following the hyperplasia, hypertrophy, and increased neovascularization of type I cells [266, 267]. HIF1α likely contributes to structural reorganization of the chemoreceptor organ through increased transcription of VEGF, receptors associated with VEGF actions (Fit-1 and Flk-1), as well as platelet derived endothelial cell growth factor expression. Flk-1 receptors regulate type I cell hyperplasia, while Fit-1 receptors promote neo-vascularization. In theory these changes are beneficial because they decrease diffusion distance between arterial blood and carotid bodies and therefore increase the speed of response s to hypoxia..

Endothelin-1 (ET-1) is a vasoactive peptide which may enhance future hypersensitivity of the carotid body following chronic hypoxia. ET-1 has been shown to phosphorylate Ca^{2+} channels, which would suggest its involvement in the potentiation of the sensory response to hypoxia [268]. ET-1 and its receptors are undetectable in normal carotid bodies, and exogenous application of ET-1 agonists to carotid bodies produces no appreciable excitatory effect. However, ET-1 expression increases under the influence of HIF1α and markedly augments the hypoxic sensory response [269]. Further support for this relationship comes from studies that demonstrate an ET-1 receptor antagonist prevents the enhanced hypoxic sensory response of neonatal carotid bodies treated with intermittent hypoxia [270]. Overall these results suggest that ET-1 may confer enhanced oxygen sensitivity through G protein-coupled ET-1 receptors to increase calcium currents [269, 271]. ET-1 may also exert its effects within the CNS through specific neuromodulator effects—mice with ET-1 heterozygous knockouts have a blunted HVR—although this hypothesis has yet to be tested [272].

HIF-1α also upregulates both HO-1 and iNOS. During chronic hypoxia, HO-2 expression decreases and the expression of the inducible form of heme oxygenase (HO-1) increases [273]. Although no evidence exists to suggest HO-1 directly contributes to an increase in CNS gain or the overall ventilatory response, HO-1 may affect the hypoxic sensitivity of carotid bodies through generating sighs and sympathetic activity [274]. It was originally proposed that iNOS is increased during prolonged hypoxic exposure mediated by HIF1α [275, 276]; however, a study by Ward et al., (2005) showed that HIF-1α may affect NO levels through transcription of nNOS [277]. As previously mentioned, increased levels of NO yield depressive effects on the oxygen sensitivity of the carotid body and thus its upregulation in hypoxia has been suggested to contribute to VAH [278]. In support of this, nNOS antagonism in the brainstem impairs the respiratory drive in chronically hypoxic mice [279].

Increasing Delivery: Vasodilation, Vasoconstriction, and Bloodborne-Oxygen Carriers

Tissue-specific responses to acute hypoxia are facilitated by the relationship between smooth muscle cells and type I cells. Hypoxic inhibition of K^+ channels in PASMCs is a well-studied event leading to an increase in intracellular $Ca^{2+,}$ which triggers the release of neurotransmitters and vasoconstriction [280, 281]. Interestingly, the response of other tissues to low oxygen stress is vasodilation as a result K_{ATP} channel opening in systemic smooth muscle cells, and consequent hyperpolarization of the membrane. Inhibiting Ca^{2+} influx into the cell causes muscle cell relaxation. These findings are consistent with studies demonstrating inhibition of the ETS radically influences oxygen sensing in NEBs and smooth muscle cells.

Organs capable of erythropoiesis (red blood cell production) respond to decreases in arterial blood oxygen levels by increasing erythrocytes and the oxygen carrying capacity of blood [282]. Abundant data shows the PHD2/pVHL/HIF2α axis stimulates this process in humans and mice. Patients with mutations in the genes encoding VHL, PHD2, and HIF2α possess elevated haemoglobin and red blood cell count s as a result of their inability to hydroxylate and ubiquitinate HIF1α proteins [283]. The resulting upregulation of HIF1α and its downstream gene targets results in congenital polycythemia, and a global upheaval of oxygen homeostasis manifested by irregular respiratory parameters and abnormal blood pressure in both normoxia and acute hypoxia [284, 285]. Additionally, mouse models possessing forced mutations in PHD2 and HIF2α exhibit impaired erythropoiesis [283].

One of the most well known responses of HIF1α to hypoxia is to increase the transcription of VEGF. In ischemic tissues, endothelial cells respond to cues in order to form new blood vessels. VEGF directs the migration of mature endothelial cells toward a hypoxic environment where they ultimately form new blood vessels in order to reoxygenate ischemic cells and tissues. Localized hypoxic stress within muscles encourages the growth of new capillaries by directing the migration of mature endothelial cells toward the region over a period of days—many genes involved in angiogenesis are targets of HIF1α including VEGF, matrix metalloproteinase 2, cathespin D, and keratin (Reviewed in [286]). Furthermore, iron is required for heme formation and is the most common limiting factor in erythropoiesis. Transferrin and ceruloplasmin /ferroxidase are both necessary in providing iron supply to erythroid tissue and are both upregulated by HIF-1α [287, 288].

Reducing Demand

One of the most common adaptations to prolonged hypoxia is a global reduction in metabolic energy expenditure associated with a coordinated reallocation of cellular resources to favour processes necessary to preserve cellular integrity [200, 235]. During periods of low oxygen stress, many organisms redirect metabolic resources away from energy-expensive functions such as growth, membrane transport, and reproduction. Organisms may also turn to anaerobic energy producing pathways to compensate for some of their energetic needs (a Pasteur effect). For most vertebrates, upregulating glycolysis is not sustainable due to the accumulation of deleterious acidic end-products; however, in the presence of sufficient glycogen and with the ability to either sequester or offload acidic end products, certain hypoxia-adapted organisms are able to reduce their metabolic rate sufficiently to rely on fermentative ATP production for prolonged periods of extreme oxygen deprivation [200, 235]. For example, hepatocytes isolated from the anoxia -tolerant turtle (*Chrysemys picta belli*) exhibit a 90% reduction in ATP-utilization [289, 290]. Metabolic suppression at the neuronal level is also critical during prolonged anoxia in these turtles [291-297], and also in goldfish [298], and this suppression is achieved primarily through a combination of channel arrest of excitatory glutamatergic receptors and spike arrest mediated by enhanced inhibitory GABAergic signaling during anoxia [299]. Previously, we have reported an ~ 85% reduction in metabolic rate in African naked mole rats exposed to hypoxic levels of 3% O_2 [300] and it is interesting to note that there is evidence of reduced excitatory activity in the brain of this hypoxia-tolerant mammal during ischemic or hypoxic stresses [301, 302], similar to previous reports in turtle and goldfish. Taken together, it is clear that surviving hypoxia requires an immediate and well-synchronized restructuring of cellular processes involved in both energy provision and utilization.

Some vertebrates amplify the biochemical changes underlying metabolic suppression by exploiting behavioural strategies that reduce thermoregulatory set points, or by entering a state of torpor. For example, the preferred temperature for freshwater fish [303, 304], reptiles [305], salamanders, and crayfish [306], have been shown to decrease as a direct result of exposure to hypoxia. Additionally, reductions in conspecific interactions (huddling) have been demonstrated in rat pups exposed to hypoxia [307]. In most vertebrates whole-body metabolic expenditure is halved for every 10°C fall in body temperature due to Q_{10} effects, offering significant savings to organisms that are able to take advantage of strategies to reduce activity and body temperature [308, 309].

CONCLUSION

Responding appropriately to changes in environmental and systemic oxygen availability is critical to maintaining cellular and systemic homeostasis in all animals. Sensitive detection of both environmental oxygen availability and changes in physiological oxygen levels, effective integration of these signals with the control of physiological function, and careful matching of metabolic demand to oxygen supply are key to the survival of all species that rely on oxygen as a terminal electron acceptor in the production of cellular energy. Vertebrates have evolved a remarkably diverse suite of adaptations throughout the oxygen transport cascade that fine-tune these systemic responses to match changes in environmental oxygen availability and/or physiological oxygen demand. These adaptations are unique and specific to each species and serve to extend the environmental range of organisms and to enhance their fitness in the face of daunting external challenges. Nonetheless, remarkable similarities exist between oxygen sensing systems in very disparate species, at the functional, cellular, and integrated organ levels.

Interestingly, HIF appears to play a major role in the hypoxia - responsiveness of most species studied to date. Indeed, HIF is present in even simple metazoans with no dedicated architecture for oxygen delivery. Therefore, it is possible that HIF arose to coordinate the switch between oxidative and glycolytic metabolism during periods of unstable oxygen conditions. Indeed, HIF1α coordinates a shift to non-oxidative forms of

ATP production by promoting the expression of genes encoding glucose transporters and nearly all known glycolytic enzymes [310, 311]. Despite our advancing knowledge in this area, considerable work remains to elucidate in particular the cellular signaling mechanisms that underlie communication between chemotransduction and the regulation of physiological responses, and also cell-level adaptations to changes in environmental and physiological oxygen availability in non-mammals, throughout development, and also during periods of physical activity.

REFERENCES

[1] Buick R. When did oxygenic photosynthesis evolve? Philosophical transactions of the Royal Society of London Series B, *Biological sciences.* 2008;363:2731-43.

[2] Raymond J, Segrè D. The effect of oxygen on biochemical networks and the evolution of complex life. *Science.* 2006;311:1764-7.

[3] Henze K, Martin W. Evolutionary biology: Essence of mitochondria. *Nature.* 2003;426:127-8.

[4] Taylor CT, McElwain JC, Beerling D, Berner R, Beerling D, Berner R, et al., Ancient atmospheres and the evolution of oxygen sensing via the hypoxia -inducible factor in metazoans. *Physiology* (Bethesda, Md). 2010;25:272-9.

[5] Oxidative stress as a mediator of life history trade -off s: Mechanisms, measurements and interpretation, (2009).

[6] Costa NA, Gut AL, Pimentel JA, Cozzolino SM, Azevedo PS, Fernandes AA, et al., Erythrocyte selenium concentration predicts intensive care unit and hospital mortality in patients with septic shock: a prospective observational study. *Crit Care.* 2014;18(3):R92.

[7] Dudley R. Atmospheric oxygen, giant Paleozoic insects and the evolution of aerial locomotor performance. *The Journal of experimental biology.* 1998;201:1043-50.

[8] Respiratory Biology: Why Insects Evolved Discontinuous Gas Exchange, (2007).

[9] Rubinow SI, Dembo M. The facilitated diffusion of oxygen by hemoglobin and myoglobin. *Biophysical journal.* 1977;18:29-42.

[10] Wittenberg JB. The molecular mechanism of hemoglobin -facilitated oxygen diffusion. *Journal of Biological Chemistry.* 1966;241:104-14.

[11] Dzal YA, Jenkin SE, Lague SL, Reichert MN, York JM, Pamenter ME. Oxygen in demand: How oxygen has shaped vertebrate physiology. *Comp Biochem Physiol A Mol Integr Physiol.* 2015.

[12] Zaccone D, Dabrowski K, Lauriano ER, De Pasquale A, Macr?? D, Satora L, et al., The simultaneous presence of neuroepithelial cells and neuroepithelial bodies in the respiratory gas bladder of the longnose gar, Lepisosteus osseus, and the spotted gar, L. oculatus. *Acta Histochemica.* 2012;114:370-8.

[13] Zaccone G, Mauceri A, Fasulo S. Neuropeptides and nitric oxide synthase in the gill and the air-breathing organs of fishes. *Journal of Experimental Zoology Part A: Comparative Experimental Biology.* 2006;305A:428-39.

[14] Burleson ML, Mercer SE, Wilk-Blaszczak MA. Isolation and characterization of putative O2 chemoreceptor cells from the gills of channel catfish (Ictalurus punctatus). *Brain Res.* 2006;1092(1):100-7.

[15] Jonz MG, Fearon IM, Nurse CA. Neuroepithelial oxygen chemoreceptors of the zebrafish gill. *J Physiol.* 2004;560(Pt 3):737-52.

[16] Zachar P, Jonz M. Characterization of ion channels an O2 sensitivity in gill neuroepithelial cells of the anoxia -tolerant goldfish (Carassius auratus). 2017.

[17] Coolidge EH, Ciuhandu CS, Milsom WK. A comparative analysis of putative oxygen -sensing cells in the fish gill. *J Exp Biol.* 2008;211 (Pt 8):1231-42.

[18] Dunel-Erb S, Bailly Y, Laurent P. Neuroepithelial cells in fish gill primary lamellae. Journal of applied physiology (Bethesda, Md: 1985). 1982;53:1342-53.

[19] Jonz MG, Nurse CA. Neuroepithelial cells and associated innervation of the zebrafish gill: A confocal immunofluorescence study. *Journal of Comparative Neurology.* 2003;461:1-17.

[20] Tzaneva V, Perry SF. The control of breathing in goldfish (Carassius auratus) experiencing thermally induced gill remodelling. *The Journal of experimental biology*. 2010;213:3666-75.

[21] Sundin L, Nilsson S. Branchial innervation2002.

[22] Burleson ML, Milsom WK. Cardio-ventilatory control in rainbow trout: II. Reflex effects of exogenous neurochemicals. *Respir Physiol*. 1995;101(3):289-99.

[23] Burleson ML, Milsom WK. Cardio-ventilatory control in rainbow trout: I. Pharmacology of branchial, oxygen -sensitive chemo receptor s. *Respir Physiol*. 1995;100(3):231-8.

[24] Milsom WK, Brill RW. Oxygen sensitive afferent information arising from the first gill arch of yellowfin tuna. *Respir Physiol*. 1986;66(2):193-203.

[25] Milsom WK. Phylogeny of respiratory chemoreceptor function in vertebrates. *Zoology-Analysis of Complex Systems*. 1998;101:316-32.

[26] Bailly Y, Dunel-Erb S, Laurent P. The neuroepithelial cells of the fish gill filament: Indolamine- immunocytochemistry and innervate on. *The Anatomical Record*. 1992;233:143-61.

[27] Milsom WK, Burleson ML. Peripheral arterial chemoreceptors and the evolution of the carotid body. *Respir Physiol Neurobiol*. 2007;157(1):4-11.

[28] Sundin L, Burleson ML, Sanchez AP, Amin-Naves J, Kinkead R, Gargaglioni LH, et al., Respiratory chemoreceptor function in vertebrates-Comparative and evolutionary aspects. *Integrative and Comparative Biology*. 2007;47:592-600.

[29] Randall D. The control of respiration and circulation in fish during exercise and hypoxia. *Journal of Experimental Biology*. 1982;100:27 5-88.

[30] Burleson ML, Smatresk NJ. Evidence for two oxygen -sensitive chemoreceptor loci in channel catfish, Ictalurus punctatus. *Physio logical Zoology*. 1990;63:208-21.

[31] McKenzie DJ, Burleson ML, Randall DJ. The effects of branchial denervation and pseudobranch ablation on cardioventilatory control in an air-breathing fish. *Journal of Experimental Biology*. 1991;161:347-65.

[32] Sundin L, Reid SG, Rantin FT, Milsom WK. Branchial receptors and cardiorespiratory reflexes in a neotropical fish, the tambaqui (Colossoma macropomum). *J Exp Biol*. 2000;203(Pt 7):1225-39.

[33] Sundin LI, Reid SG, Kalinin AL, Rantin FT, Milsom WK. Cardiovascular and respiratory reflexes: the tropical fish, traira (Hoplias malabaricus) O2 chemoresponses. *Respir Physiol.* 1999;116(2-3):181-99.

[34] Milsom WK, Reid SG, Rantin FT, Sundin L. Extrabranchial chemoreceptors involved in respiratory reflexes in the neotropical fish Colossoma macropomum (the tambaqui). *J Exp Biol.* 2002;2 05(Pt 12):1765-74.

[35] McKenzie DC, Goodman LS, Nath C, Davidson B, Matheson GO, Parkhouse WS, et al., Cardiovascular adaptations in Andean natives after 6 wk of exposure to sea level. *J Appl Physiol.* 1991;70(6):2650-5.

[36] Laurent P. Gill Internal Morphology. Gills Anatomy, Gas Transfer, and Acid-Base Regulation. 1984;10, Part A:73-183.

[37] Neil E, Ström L, Zotterman Y. Action Potential Studies of Afferent Fibres in the IXth and Xth Cranial Nerves of the Frog. *Acta Physiologica Scandinavica.* 1950;20:338-50.

[38] Van Vliet BN, West NH. Functional characteristics of arterial chemoreceptors in an amphibia n (Bufo marinus). *Resp Physiol.* 1992;88:113-27.

[39] Kusakabe T. Carotid labyrinth of amphibians. *Microscopy Research and Technique.* 2002;59:207-26.

[40] Jonz MG, Nurse CA. Peripheral Chemoreceptors in Air- Versus Water- Breathers. *Advances in experimental medicine and biology* 2012. p. 19-27.

[41] Abdel-Magied EM, King aS. The topographical anatomy and blood supply of the carotid body region of the domestic fowl. *Journal of anatomy*. 1978;126:535-46.

[42] Kameda Y, Miura M, Hayashida Y. Different effects of prolonged isocapnic hypoxia on the carotid body and the glomus cells in the wall of the common carotid artery of the chicken. *Brain Research*. 1998;805:191-206.

[43] Reichert MN, Brink DL, Milsom WK. Evidence for a carotid body homolog in the lizard Tupinambis merianae. *Journal of Experimental Biology*. 2015;218.

[44] Reyes C, Fong AY, Milsom WK. Distribution and innervation of putative peripheral arterial chemoreceptors in the red-eared slider (Trachemys scripta elegans). *Journal of Comparative Neurology*. 2015;523:1399-418.

[45] Jia X, Burggren W. Developmental changes in chemoreceptive control of gill ventilation in larval bullfrogs (Rana catesbeiana). II. Sites of O2-sensitive chemoreceptors. *J Exp Biol*. 1997;200 (Pt 16):2237-48.

[46] West NH, Langille BL, Jones DR. Cardiovascular System. In: King AS, McLelland, J., editor. Form and Function In Birds. London: Academic Press; 1981.

[47] Crank WD, Kuhlmann WD, Fedde MR. Functional localization of avian intrapulmonary CO2 receptors within the parabronchial mantle. *Resp Physiol*. 1980;41:71-85.

[48] Hempleman SC, Posner RG. CO_2 transduction mechanisms in avian intrapulmonary chemoreceptors: experiments and models. *Respir Physiol Neurobiol*. 2004;144(2-3):203-14.

[49] Peers C. Oxygen-sensitive ion channels. *Trends in Pharmacological Sciences*. 1997;18:405-8.

[50] Lopez-Barneo J, Pardal R, Ortega-Sáenz P. Cellular mechanism of oxygen sensing. *Annual review of physiology*. 2001;63:259-87.

[51] Fu XW, Nurse CA, Wong V, Cutz E. Hypoxia-induced secretion of serotonin from intact pulmonary neuroepithelial bodies in neonatal rabbit. *The Journal of physiology*. 2002;539:503-10.

[52] Youngson C, Nurse C, Yeger H, Cutz E. Oxygen sensing in airway chemoreceptors. *Nature.* 1993;365:153-5.

[53] Fitzgerald RS, Lahiri S. Reflex responses to chemoreceptor stimulation. Handbook of Physiology, Sect 3: The respiratory system, *Vol II: Control of breathing, Part 2.* 1986:313-62.

[54] Stahlman MT, Jones M, Gray ME, Kasselberg AG, Vaughn WK. Ontogeny of neuroendocrine cells in human fetal lung. III. An electron microscopic immunohistochemical study. *Laboratory investigation; a journal of technical methods and pathology.* 1987;5 6:629-41.

[55] Fu XW, Nurse CA, Wang YT, Cutz E. Selective modulation of membrane currents by hypoxia in intact airway chemoreceptors from neonatal rabbit. *Journal of Physiology.* 1999;514:139-50.

[56] O'Kelly I. O2 Sensing by Airway Chemoreceptor-derived Cells. Protein Kinase C Activation Reveals Functional Evidence for Involvement of NADPH Oxidase. *Journal of Biological Chemistry.* 2000;275:7684-92.

[57] O'Kelly I, Peers C, Kemp PJ. O2-sensitive K+ channels in neuroepithelial body-derived small cell carcinoma cells of the human lung. *The American journal of physiology.* 1998;275:L709-16.

[58] Prabhakar NR. O_2 sensing at the mammalian carotid body: why multiple O_2 sensors and multiple transmitters? *Exp Physiol.* 2006;91(1):17-23.

[59] Kumar P, Prabhakar NR. Peripheral chemoreceptors: function and plasticity of the carotid body. *Compr Physiol.* 2012;2(1):141-219.

[60] Kim DK, Prabhakar NR, Kumar GK. Acetylcholine release from the carotid body by hypoxia: evidence for the involvement of autoinhibitory receptors. *J Appl Physiol* (1985). 2004;96(1):376-83.

[61] Kim DK, Summers BA, Prabhakar NR, Kumar GK. Neurotransmitter release from the rabbit carotid body: differential effects of hypoxia on substance P and acetylcholine release. *Adv Exp Med Biol.* 2001;499:39-43.

[62] Prabhakar NR. Neurotransmitters in the carotid body. *Advances in Experimental Medicine and Biology.* 1994;360:57-69.

[63] Fidone SJ, Gonzalez C. Initiation and Control of Chemoreceptor Activity in the Carotid Body. *Handbook of Physiology, The Respiratory System, Control of Breathing.* 2011.
[64] Signal processing at mammalian carotid body chemoreceptors, (2013).
[65] Gonzalez C, Almaraz L, Obeso A, Rigual R. Carotid body chemoreceptors: from natural stimuli to sensory discharges. *Physiol Rev.* 1994;74(4):829-98.
[66] Tse A, Yan L, Lee AK, Tse FW. Autocrine and paracrine actions of ATP in rat carotid body. *Canadian Journal of Physiology and Pharmacology.* 2012;90:705-11.
[67] Nurse CA. Synaptic and paracrine mechanisms at carotid body arterial chemoreceptors. *The Journal of physiology.* 2014;16:3419-26.
[68] Transduction of chemostimuli by the type I carotid body cell, (2004).
[69] Biscoe TJ, Purves MJ, Sampson SR. The frequency of nerve impulses in single carotid body chemoreceptor afferent fibres recorded in vivo with intact circulation. *The Journal of physiology.* 1970;208:121-31.
[70] Kumar GK, Prabhakar NR. Tachykinins in the control of breathing by hypoxia: pre- and post-genomic era. *Respir Physiol Neurobiol.* 2003;135(2-3):145-54.
[71] CO_2, brainstem chemoreceptors and breathing, (1999).
[72] Prabhakar NR. Oxygen sensing by carotid body chemoreceptors. *Journal of Applied Physiology*. 2000;88:2287-95.
[73] López-Barneo J, López-López JR, Ureña J, González C. Chemotransduction in the carotid body: K+ current modulated by PO2 in type I chemoreceptor cells. *Science*. 1988;241:580-2.
[74] Duchen MR, Caddy KW, Kirby GC, Patterson DL, Ponte J, Biscoe TJ. Biophysical studies of the cellular elements of the rabbit carotid body. *Neuroscience.* 1988;26:291-311.
[75] Lopez-Barneo J, Lopez-Lopez JR, Urena J, Gonzalez C. Chemotransduction in the carotid body: K+ current modulated by PO2 in type I chemoreceptor cells. *Science.* 1988;241(4865):580-2.

[76] Hescheler J, Delpiano MA, Acker H, Pietruschka F. Ionic currents on type-I cells of the rabbit carotid body measured by voltage-clamp experiments and the effect of hypoxia. *Brain Research.* 1989;486:79-88.

[77] Ureña J, López-López J, González C, López-Barneo J. Ionic currents in dispersed chemoreceptor cells of the mammalian carotid body. *The Journal of general physiology.* 1989;93:979-99.

[78] Peers C. Hypoxic suppression of K+ currents in type I carotid body cells: Selective effect on the Ca2+ -activated K+ current. *Neuro science Letters.* 1990;119:253-6.

[79] Stea A, Nurse CA. Whole-cell and perforated-patch recordings from O2-sensitive rat carotid body cells grown in short- and long-term culture. *Pflugers Arch.* 1991;418(1-2):93-101.

[80] Shirahata M, Sham JS. Roles of ion channels in carotid body chemotransmission of acute hypoxia. *The Japanese journal of physiology.* 1999;49:213-28.

[81] Gu XQ, Pamenter ME, Siemen D, Sun X, Haddad GG. Mitochondrial but not plasmalemmal BK channels are hypoxia - sensitive in human glioma. *Glia.* 2014.

[82] Wasicko MJ, Breitwieser GE, Kim I, Carroll JL. Postnatal development of carotid body glomus cell response to hypoxia. *Resp Physiol Neurobi.* 2006;154:356-71.

[83] Kemp PJ. Detecting acute changes in oxygen: will the real sensor please stand up? *Experimental Physiology.* 2006;91:829-34.

[84] Kemp PJ, Peers C. Oxygen sensing by ion channels. *Essays in biochemistry.* 2007;43:77-90.

[85] Olschewski A, Hong Z, Nelson DP, Weir EK. Graded response of K+ current, membrane potential, and [Ca2+]i to hypoxia in pulmonary arterial smooth muscle. *Am J Physiol Lung Cell Mol Physiol.* 2002;283:L1143-L50.

[86] Lopezlopez JR, Gonzalez C. Time Course of K+ Current Inhibition By Low Oxygen in Chemoreceptor Cells of Adult-Rabbit Carotid-Body-Effects of Carbon-Monoxide. *FEBS Lett.* 1992;299:251-4.

[87] Qin Z, Lewis JE, Perry SF. Zebrafish (Danio rerio) gill neuroepithelial cells are sensitive chemoreceptors for environmental CO_2. *The Journal of Physiology*. 2010;588:861-72.

[88] Buckler KJ. TASK-like potassium channels and oxygen sensing in the carotid body. *Respir Physiol Neurobiol*. 2007;157(1):55-64.

[89] Wang R. Two's company, three's a crowd: can H2S be the third endogenous gaseous transmitter? *FASEB journal: official publication of the Federation of American Societies for Experimental Biology*. 2002;16:1792-8.

[90] Moncada S, Palmer RMJ, Higgs EA. Nitric oxide: physiology, pathophysiology, and pharmacology. *Pharmacological Reviews*. 1991;43:109-42.

[91] Olson KR, Donald JA, Dombkowski RA, Perry SF. Evolutionary and comparative aspects of nitric oxide, carbon monoxide and hydrogen sulfide. *Resp Physiol Neurobi*. 2012;184:117-29.

[92] Wang ZZ, Stensaas LJ, Bredt DS, Dinger B, Fidone SJ. Localization and actions of nitric oxide in the cat carotid body. *Neuroscience*. 1994;60:275-86.

[93] Endogenous carbon monoxide in control of respiration, (1998).

[94] Prabhakar NR. Carbon monoxide (CO) and hydrogen sulfide (H(2)S) in hypoxic sensing by the carotid body. *Respir Physiol Neurobiol*. 2012;184(2):165-9.

[95] Prabhakar NR, Semenza GL. Gaseous messengers in oxygen sensing. *J Mol Med* (Berl). 2012;90(3):265-72.

[96] Olson KR. Hydrogen sulfide and oxygen sensing: implications in cardiorespiratory control. *The Journal of experimental biology*. 2008;211:2727-34.

[97] Porteus CS, Abdallah SJ, Pollack J, Kumai Y, Kwong RW, Yew HM, et al., The role of hydrogen sulphide in the control of breathing in hypoxic zebrafish (Danio rerio). *J Physiol*. 2014.

[98] Williams SEJ, Wootton P, Mason HS, Bould J, Iles DE, Riccardi D, et al., Hemoxygenase-2 is an oxygen sensor for a calcium -sensitive potassium channel. *Science* (New York, NY). 2004;306:2093-7.

[99] Adachi T, Ishikawa K, Hida W, Matsumoto H, Masuda T, Date F, et al., Hypoxemia and blunted hypoxic ventilatory responses in mice lacking heme oxygenase -2. *Biochem Biophys Res Commun.* 2004;320(2):514-22.

[100] Ortega-Sáenz P, Pascual A, Gómez-Díaz R, López-Barneo J. Acute oxygen sensing in heme oxygenase -2 null mice. *The Journal of general physiology.* 2006;128:405-11.

[101] Wang D, Zhong X-P, Qiao Z-X, Gui J-F. Inductive transcription and protective role of fish heme oxygenase -1 under hypoxic stress. *The Journal of experimental biology.* 2008;211:2700-6.

[102] Tzaneva V, Perry SF. Heme oxygenase-1 (HO-1) mediated respiratory responses to hypoxia in the goldfish, Carassius auratus. *Resp Physiol Neurobi.* 2014;199:1-8.

[103] Prevot-D'Alvise N, Richard S, Coupé S, Bunet R, Grillasca JP. Acute toxicity of a commercial glyphosate formulation on European sea bass juveniles (Dicentrarchus labrax L.): gene expression s of heme oxygenase -1 (ho-1), acetylcholinesterase (AChE) and aromatases (cyp19a and cyp19b). *Cellular and molecular biology* (Noisy-le-Grand, France). 2013;59.

[104] Li FJ, Duggal RN, Oliva OM, Karki S, Surolia R, Wang Z, et al., Heme oxygenase-1 protects Corexit 9500ainduced respiratory epithelial injury across species. *PLoS ONE.* 2015;10.

[105] Prabhakar NR, Kumar GK, Chang CH, Agani FH, Haxhiu MA. Nitric oxide in the sensory function of the carotid body. *Brain Res.* 1993;625(1):16-22.

[106] Summers BA, Overholt JL, Prabhakar NR. Nitric oxide inhibits L-type Ca2+ current in glomus cells of the rabbit carotid body via a cGMP-independent mechanism. *J Neurophysiol.* 1999;81(4):1449-57.

[107] Rengasamy A, Johns RA. Characterization of endothelium -derived relaxing factor/nitric oxide synthase from bovine cerebellum and mechanism of modulation by high and low oxygen tensions. *The Journal of pharmacology and experimental therapeutics.* 1991;259:310-6.

[108] Kline DD, Yang T, Huang PL, Prabhakar NR. Altered respiratory responses to hypoxia in mutant mice deficient in neuronal nitric oxide synthase. *J Physiol.* 1998;511 (Pt 1):273-87.

[109] Hyndman Ka, Choe KP, Havird JC, Rose RE, Piermarini PM, Evans DH. Neuronal nitric oxide synthase in the gill of the killifish, Fundulus heteroclitus. *Comparative biochemistry and physiology Part B, Biochemistry & molecular biology.* 2006;144:510-9.

[110] Wang W-S, Hung S-W, Lin Y-H, Tu C-Y, Wong M-L, Chiou S-H, et al., Purification and Localization of Nitric Oxide Synthases from Hybrid Tilapia (Nile Tilapia × Mozambique Tilapia). *Journal of Aquatic Animal Health.* 2007;19:168-78.

[111] Porteus CS, Pollack J, Tzaneva V, Kwong RWM, Kumai Y, Abdallah SJ, et al., A role for nitric oxide in the control of breathing in zebrafish (Danio rerio). *Journal of Experimental Biology.* 2015;218:3746-53.

[112] Dorman DC, Moulin FJM, McManus BE, Mahle KC, James RA, Struve MF. Cytochrome oxidase inhibition induced by acute hydrogen sulfide inhalation: Correlation with tissue sulfide concentrations in the rat brain, liver, lung, and nasal epithelium. *Toxicological Sciences.* 2002;65:18-25.

[113] The inhibition of mitochondrial cytochrome oxidase by the gases carbon monoxide, nitric oxide, hydrogen cyanide and hydrogen sulfide: *Chemical mechanism and physiological significance,* (2008).

[114] Perry SF, McNeill B, Elia E, Nagpal A, Vulesevic B. Hydrogen sulfide stimulates catecholamine secretion in rainbow trout (Oncorhynchus mykiss). *American journal of physiology Regulatory, integrative and comparative physiology.* 2009;296:R133-R40.

[115] Olson KR, Dombkowski Ra, Russell MJ, Doellman MM, Head SK, Whitfield NL, et al., Hydrogen sulfide as an oxygen sensor /transducer in vertebrate hypoxic vasoconstriction and hypoxic vasodilation. *The Journal of experimental biology.* 2006;209:4011-23.

[116] Gasotransmitters: Novel regulators of epithelial Na + transport ?, (2012).

[117] Peng YJ, Nanduri J, Raghuraman G, Souvannakitti D, Gadalla MM, Kumar GK, et al., H2S mediates O_2 sensing in the carotid body. *Proc Natl Acad Sci U S A.* 2010;107(23):10719-24.

[118] Telezhkin V, Brazier SP, Cayzac SH, Wilkinson WJ, Riccardi D, Kemp PJ. Mechanism of inhibition by hydrogen sulfide of native and recombinant BKCa channels. *Resp Physiol Neurobi.* 2010;172:169-78.

[119] Nicholls P, Kim J. Sulphide as an inhibitor and electron donor for the cytochrome c oxidase system. *Canadian journal of biochemistry.* 1982;60:613-23.

[120] Peng Y-J, Nanduri J, Raghuraman G, Souvannakitti D, Gadalla MM, Kumar GK, et al., H2S mediates O2 sensing in the carotid body. *P Natl Acad Sci USA.* 2010;107:10719-24.

[121] Makarenko VV, Nanduri J, Raghuraman G, Fox AP, Gadalla MM, Kumar GK, et al., Endogenous H2S is required for hypoxic sensing by carotid body glomus cells. *Am J Physiol Cell Physiol.* 2012;303(9):C916-23.

[122] Olson KR, Healy MJ, Qin Z, Skovgaard N, Vulesevic B, Duff DW, et al., Hydrogen sulfide as an oxygen sensor in trout gill chemoreceptors. *American journal of physiology Regulatory, integrative and comparative physiology.* 2008;295:R669-80.

[123] Fedde MR, Scheid P. Intrapulmonary CO_2 receptors in the duck: IV. Discharge pattern of the population during a respiratory cycle. *Resp Physiol.* 1976;26:223-7.

[124] Li Q, Sun B, Wang X, Jin Z, Zhou Y, Dong L, et al., A crucial role for hydrogen sulfide in oxygen sensing via modulating large conductance calcium -activated potassium channels. *Antioxid Redox Sign.* 2010;12:1179-89.

[125] Mustafa AK, Gadalla MM, Sen N, Kim S, Mu W, Gazi SK, et al., H2S signals through protein S-sulfhydration. *Science signaling.* 2009;2:ra72.

[126] Buckler KJ. Effects of exogenous hydrogen sulphide on calcium signaling, background (TASK) K channel activity and mitochondrial

function in chemoreceptor cells. *Pflugers Archiv European Journal of Physiology.* 2012;463:743-54.

[127] Tang G, Wu L, Liang W, Wang R. Direct stimulation of K(ATP) channels by exogenous and endogenous hydrogen sulfide in vascular smooth muscle cells. *Molecular pharmacology.* 2005;68:1757-64.

[128] Yuan G, Vasavda C, Peng Y-J, Makarenko VV, Raghuraman G, Nanduri J, et al., Protein kinase G-regulated production of H2S governs oxygen sensing. *Science signaling.* 2015;8:ra37.

[129] Consensus sequences as substrate specificity determinants for protein kinases and protein phosphatases, (1991).

[130] Ellsworth ML, Forrester T, Ellis CG, Dietrich HH. The erythrocyte as a regulator of vascular tone. *Am J Physiol.* 1995;269:H2155-61.

[131] Cosby K, Partovi KS, Crawford JH, Patel RP, Reiter CD, Martyr S, et al., Nitrite reduction to nitric oxide by deoxyhemoglobin vasodilates the human circulation. *Nature medicine.* 2003;9:1498-505.

[132] Huang Z, Shiva S, Kim-Shapiro DB, Patel RP, Ringwood LA, Irby CE, et al., Enzymatic function of hemoglobin as a nitrite reductase that produces NO under allosteric control. *The Journal of clinical investigation.* 2005;115:2099-107.

[133] Hendgen-Cotta UB, Merx MW, Shiva S, Schmitz J, Becher S, Klare JP, et al., Nitrite reductase activity of myoglobin regulates respiration and cellular viability in myocardial ischemia -reperfusion injury. *P Natl Acad Sci USA.* 2008;105:10256-61.

[134] Ignarro LJ, Buga GM, Wood KS, Byrns RE, Chaudhurit G, Sawyer CH. Endothelium-derived relaxing factor produced and released from artery and vein is nitric oxide (endothelium -dependent relaxation / vascular smooth muscle /cyclic GMP). *Medical Sciences.* 1987;84:9265-9.

[135] Vascular endothelial cells synthesize nitric oxide from L-arginine., (1988).

[136] Nagababu E, Ramasamy S, Abernethy DR, Rifkind JM. Active Nitric Oxide Produced in the Red Cell under Hypoxic Conditions by

Deoxyhemoglobin-mediated Nitrite Reduction. *Journal of Biological Chemistry.* 2003;278:46349-56.

[137] Crawford JH, Isbell TS, Huang Z, Shiva S, Chacko BK, Schechter AN, et al., Hypoxia, red blood cells, and nitrite regulate NO-dependent hypoxic vasodilation. *Blood.* 2006;107:566-74.

[138] Jensen FB. The dual roles of red blood cells in tissue oxygen delivery: oxygen carriers and regulators of local blood flow. The Journal of experimental biology. 2009;212:3387-93.

[139] Pamenter ME, Ali SS, Tang Q, Finley JC, Gu XQ, Dugan LL, et al., An in vitro ischemic penumbral mimic perfusate increases NADPH oxidase-mediated superoxide production in cultured hippocampal neurons. *Brain Res*. 2012;1452:165-72.

[140] Forman HJ. Signal transduction and reactive species. *Free Radic Biol Med.* 2009;47(9):1237-8.

[141] Mittal M, Gu XQ, Pak O, Pamenter ME, Haag D, Fuchs DB, et al., Hypoxia induces Kv channel current inhibition by increased NADPH oxidase-derived reactive oxygen species. *Free Radic Biol Med.* 2012;52(6):1033-42.

[142] Griendling KK, Sorescu D, Ushio-Fukai M. NAD (P)H oxidase: role in cardiovascular biology and disease. *Circulation research.* 2000;86:494-501.

[143] Kummer W, Acker H. Immunohistochemical demonstration of four subunits of neutrophil NAD (P)H oxidase in type I cells of carotid body. *Journal of applied physiology.* 1995;78:1904-9.

[144] Marshall C, Mamary AJ, Verhoeven AJ, Marshall BE. Pulmonary Artery NADPH-oxidase Is Activated in Hypoxic Pulmonary Vasoconstriction. *American Journal of Respiratory Cell and Molecular Biology.* 1996;15:633-44.

[145] Archer SL, Reeve HL, Michelakis E, Puttagunta L, Waite R, Nelson DP, et al., O2 sensing is preserved in mice lacking the gp91 phox subunit of NADPH oxidase. *Proceedings of the National Academy of Sciences.* 1999;96:7944-9.

[146] Roy A, Rozanov C, Mokashi A, Daudu P, Al-mehdi AB, Shams H, et al., Mice lacking in gp91 phox subunit of NAD (P)H oxidase showed

glomus cell [Ca(2+)](i) and respiratory responses to hypoxia. *Brain research.* 2000;872:188-93.

[147] Kazemian P, Stephenson R, Yeger H, Cutz E. Respiratory control in neonatal mice with NADPH oxidase deficiency. *Respir Physiol.* 2001;126:89-101.

[148] Obeso A, Gomez-Nino A, Gonzalez C. NADPH oxidase inhibition does not interfere with low PO2 transduction in rat and rabbit CB chemoreceptor cells. *Am J Physiol. 1999*;276:C593-601.

[149] Sanders KA, Sundar KM, He L, Dinger B, Fidone S, Hoidal JR. Role of components of the phagocytic NADPH oxidase in oxygen sensing. *J Appl Physiol.* 2002;93:1357-64.

[150] Fearon IM, Varadi G, Koch S, Isaacsohn I, Ball SG, Peers C. Splice variants reveal the region involved in oxygen sensing by recombinant human L-type Ca(2+) channels. *Circulation research.* 2000;87:537-9.

[151] Franco-Obregón A, Ureña J, López-Barneo J. Oxygen-sensitive calcium channels in vascular smooth muscle and their possible role in hypoxic arterial relaxation. *P Natl Acad Sci USA.* 1995;92:4715-9.

[152] Xiong Z, Sperelakis N. Regulation of L-type calcium channels of vascular smooth muscle cells. *Journal of Molecular and Cellular Cardiology.* 1995;27:75-91.

[153] Gollasch M, Löhn M, Fürstenau M, Nelson MT, Luft FC, Haller H. Ca2+ channels, Ca2+ sparks, and regulation of arterial smooth muscle function. *Zeitschrift fur Kardiologie.* 2000;89 Suppl 2:15-9.

[154] Hirenallur SD, Haworth ST, Leming JT, Chang J, Hernandez G, Gordon JB, et al., Upregulation of vascular calcium channels in neonatal piglets with hypoxia -induced pulmonary hypertension. *Am J Physiol Lung Cell Mol Physiol.* 2008;295:L915-24.

[155] Fieber LA, McCleskey EW. L-type calcium channels in type I cells of the rat carotid body. *J Neurophysiol.* 1993;70:1378-84.

[156] e Silva MJ, Lewis DL. L- and N-type Ca2+ channels in adult rat carotid body chemoreceptor type I cells. *The Journal of physiology.* 1995;489 (Pt 3:689-99.

[157] Peers C, Carpenter E, Hatton CJ, Wyatt CN, Bee D. Ca2+ channel currents in type I carotid body cells of normoxic and chronically hypoxic neonatal rats. *Brain Research.* 1996;739:251-7.

[158] Carpenter E, Peers C. A standing Na+ conductance in rat carotid body type I cells. *Neuroreport.* 2001;12:1421-5.

[159] Pamenter ME. Mitochondria: a multimodal hub of hypoxia tolerance. *Can J Zool.* 2014;92(7):569-89.

[160] Mulligan E, Lahiri S. Dependence of carotid chemoreceptor stimulation by metabolic agents on PaO_2 and $PaCO_2$. *Journal of applied physiology: respiratory, environmental and exercise physiology.* 1981;50:884-91.

[161] Biscoe TJ, Duchen MR. Responses of type I cells dissociated from the rabbit carotid body to hypoxia. *The Journal of physiology.* 1990;428:39-59.

[162] Mosqueira M, Iturriaga R. Carotid body chemosensory excitation induced by nitric oxide: involvement of oxidative metabolism. *Resp Physiol Neurobi.* 2002;131:175-87.

[163] Wyatt CN, Buckler KJ. The effect of mitochondrial inhibitors on membrane currents in isolated neonatal rat carotid body type I cells. *J Physiol.* 2004;5561:175-91.

[164] Jones DP, Mason HS. Gradients of O2 concentration in hepatocytes. *The Journal of biological chemistry.* 1978;253:4874-80.

[165] Wilson DF, Erecinska M, Drown C, Silver IA. The oxygen dependence of cellular energy metabolism. *Arch Biochem Biophys.* 1979;195(2):485-93.

[166] Kennedy FG, Jones DP. Oxygen dependence of mitochondrial function in isolated rat cardiac myocytes. *American Journal of Physiology-Cell Physiology.* 1986;250:C374-C83.

[167] Duchen MR, Biscoe TJ. Relative mitochondrial membrane potential and [Ca2+]i in type I cells isolated from the rabbit carotid body. *The Journal of Physiology.* 1992;450:33-61.

[168] Streller T, Huckstorf C, Pfeiffer C, Acker H. Unusual cytochrome a592 with low PO_2 affinity correlates as putative oxygen sensor with rat carotid body chemoreceptor discharge. *The FASEB journal:*

official publication of the Federation of American Societies for Experimental Biology. 2002;16:1277-9.

[169] Wilson DFDF, Rumsey WLWL, Green TJTJ, Vanderkooi JMJM. The oxygen dependence of mitochondrial oxidative phosphorylation measured by a new optical method for measuring oxygen concentration. *The Journal of Biological Chemistry*. 1988;263:2712-8.

[170] Chandel N, Budinger GR, Kemp RA, Schumacker PT. Inhibition of cytochrome -c oxidase activity during prolonged hypoxia. *Am J Physiol*. 1995;268(6 Pt 1):L918-25.

[171] Chandel NS, Budinger GRS, Schumacker PT. Molecular oxygen modulates cytochrome c oxidase function. *Journal of Biological Chemistry*. 1996;271:18672-7.

[172] Biscoe TJ, Duchen MR, Eisner DA, O 'neill SC, Valdeolmillos M. Measurements of Intracellular Ca2l in Dissociated Type I Cells of the Rabbit Carotid Body. *Journal of Physiology*. 1989;416:421-34.

[173] Tamura M, Hazeki O, Nioka S, Chance B. In Vivo Study of Tissue Oxygen Metabolism using Optical and Nuclear Magnetic Resonance Spectroscopies. *Annual Review of Physiology*. 1989;51:813-34.

[174] Palacios-Callender M, Quintero M, Hollis VS, Springett RJ, Moncada S. Endogenous NO regulates superoxide production at low oxygen concentrations by modifying the redox state of cytochrome c oxidase. *Proc Natl Acad Sci U S A*. 2004;101:7630-5.

[175] Nitric oxide, cytochrome c oxidase, and the cellular response to hypoxia, (2010).

[176] Chandel NS, McClintock DS, Feliciano CE, Wood TM, Melendez JA, Rodriguez AM, et al., Reactive oxygen species generated at mitochondrial complex III stabilize hypoxia -inducible factor -1alpha during hypoxia: a mechanism of O_2 sensing. *J Biol Chem*. 2000;275(33):25130-8.

[177] Turrens JF. Mitochondrial formation of reactive oxygen species. J Physiol. 2003;552(Pt 2):335-44.

[178] Klimova T, Chandel NS. Mitochondrial complex III regulates hypoxic activation of HIF. *Cell Death Differ*. 2008;15(4):660-6.

[179] Pamenter ME, Hogg DW, Ormond J, Shin DS, Woodin MA, Buck LT. Endogenous GABA (A) and GABA(B) receptor -mediated electrical suppression is critical to neuronal anoxia tolerance. *Proc Natl Acad Sci U S A.* 2011;108(27):11274-9.

[180] Hogg DW, Pamenter ME, Dukoff DJ, Buck LT. Decreases in mitochondrial reactive oxygen species initiate GABA (A) receptor -mediated electrical suppression in anoxia -tolerant turtle neurons. *J Physiol.* 2015;593(10):2311-26.

[181] Buck LT, Hogg DW, Rodgers-Garlick C, Pamenter ME. Oxygen sensitive synaptic neurotransmission in anoxia -tolerant turtle cerebrocortex. *Adv Exp Med Biol.* 2012;In Press.

[182] Buck LT, Hogg DW, Rodgers-Garlick C, Pamenter ME. The Painted Turtle as a model of natural anoxia tolerance: role of the neurotransmitter GABA. In: MJ Cosgrove SR, editor. Turtles: Anatomy, Ecology and Conservation. New York: Nova Science; 2012.

[183] Pamenter ME, Buck LT. Neuronal membrane potential is mildly depolarized in the anoxic turtle cortex. *Comp Biochem Physiol A Mol Integr Physiol.* 2008;150(4):410-4.

[184] Milton SL, Nayak G, Kesaraju S, Kara L, Prentice HM. Suppression of reactive oxygen species production enhances neuronal survival in vitro and in vivo in the anoxia -tolerant turtle *Trachemys scripta. J Neurochem.* 2007;101(4):993-1001.

[185] Pamenter ME, Richards MD, Buck LT. Anoxia-induced changes in reactive oxygen species and cyclic nucleotides in the painted turtle. *J Comp Physiol [B].* 2007;177(4):473-81.

[186] Duerr JM, Podrabsky JE. Mitochondrial physiology of diapausing and developing embryos of the annual killifish Austrofundulus limnaeus: implications for extreme anoxia tolerance. *J Comp Physiol B.* 2010;180(7):991-1003.

[187] Hickey AJ, Renshaw GM, Speers-Roesch B, Richards JG, Wang Y, Farrell AP, et al., A radical approach to beating hypoxia: depressed free radical release from heart fibres of the hypoxia-tolerant epaulette

shark (Hemiscyllum ocellatum). *J Comp Physiol B*. 2012;182(1):91-100.

[188] Ali SS, Hsiao M, Zhao HW, Dugan LL, Haddad GG, Zhou D. Hypoxia-adaptation involves mitochondrial metabolic depression and decreased ROS leakage. *PLoS One*. 2012;7(5):e36801.

[189] Chandel NS, Schumacker PT. Cellular oxygen sensing by mitochondria: old questions, new insight. *J Appl Physiol.* 2000;8 8(5):1880-9.

[190] Mitochondrial reactive oxygen species regulate cellular signaling and dictate biological outcomes, (2010).

[191] Guzy RD, Hoyos B, Robin E, Chen H, Liu L, Mansfield KD, et al., Mitochondrial complex III is required for hypoxia -induced ROS production and cellular oxygen sensing. *Cell Metab.* 2005;1(6):401-8.

[192] Nanduri J, Wang N, Yuan G, Khan SA, Souvannakitti D, Peng Y-J, et al., Intermittent hypoxia degrades HIF-2alpha via calpains resulting in oxidative stress: implications for recurrent apnea - induced morbidities. *P Natl Acad Sci USA.* 2009;106:1199-204.

[193] Peng Y-J, Nanduri J, Yuan G, Wang N, Deneris E, Pendyala S, et al., NADPH oxidase is required for the sensory plasticity of the carotid body by chronic intermittent hypoxia. *The Journal of neuroscience: the official journal of the Society for Neuroscience.* 2009;29:4903-10.

[194] Hoshi T, Heinemann S. Regulation of cell function by methionine oxidation and reduction. *The Journal of physiology*. 2001;531:1-11.

[195] Finkel T. Signal transduction by reactive oxygen species. *J Cell Biol.* 2011;194(1):7-15.

[196] Finkel T. From sulfenylation to sulfhydration: what a thiolate needs to tolerate. *Science signaling*. 2012;5:pe10.

[197] Bernardini A, Brockmeier U, Metzen E, Berchner-Pfannschmidt U, Harde E, Acker-Palmer A, et al., Measurement of ros levels and membrane potential dynamics in the intact carotid body ex vivo. *Advances in Experimental Medicine and Biology*. 2015;860:55-9.

[198] Bernardini A, Brockmeier U, Metzen E, Berchner-Pfannschmidt U, Harde E, Acker-Palmer A, et al., Type I cell ROS kinetics under

hypoxia in the intact mouse carotid body ex vivo: a FRET based study. *American journal of physiology Cell physiology.* 2015;308:C61-7.

[199] Lindsley JE, Rutter J. Nutrient sensing and metabolic decisions. *Comparative biochemistry and physiology Part B, Biochemistry & molecular biology.* 2004;139:543-59.

[200] Buck LT, Pamenter ME. Adaptive responses of vertebrate neurons to anoxia -Matching supply to demand. *Respir Physiol Neurobiol.* 2006;154(1-2):226-40.

[201] Emerling BM, Weinberg F, Snyder C, Burgess Z, Mutlu GM, Viollet B, et al., Hypoxic activation of AMPK is dependent on mitochondrial ROS but independent of an increase in AMP/ATP ratio. *Free Radical Bio Med.* 2009;46:1386-91.

[202] Hardie DG. Minireview: the AMP-activated protein kinase cascade: the key sensor of cellular energy status. *Endocrinology.* 2003;144(12):5179-83.

[203] Evans AM, Hardie DG, Peers C, Wyatt CN, Viollet B, Kumar P, et al., Ion Channel Regulation by AMPK. *Annals of the New York Academy of Sciences.* 2009;1177:89-100.

[204] Evans aM, Mustard KJW, Wyatt CN, Peers C, Dipp M, Kumar P, et al., Does AMP-activated protein kinase couple inhibition of mitochondrial oxidative phosphorylation by hypoxia to calcium signaling in O2-sensing cells? *The Journal of biological chemistry.* 2005;280:41504-11.

[205] Evans ZP, Ellett JD, Fariss MW, Schnellmann RG, Schmidt MG, Chavin K. Vitamin E succinate reduces ischemia /reperfusion injury in steatotic livers. *Transplant Proc.* 2008;40(10):3327-9.

[206] Wyatt CN, Mustard KJ, Pearson SA, Dallas ML, Atkinson L, Kumar P, et al., AMP-activated protein kinase mediates carotid body excitation by hypoxia. *The Journal of biological chemistry.* 2007;282:8092-8.

[207] Marsin A-S, Bertrand† L, Rider MH, Deprez J, Beauloye C, Vincent‡ MF, et al., Phosphorylation and activation of heart PFK-2

by AMPK has a role in the stimulation of glycolysis during ischaemia. *Current Biology.* 2000;10:1247-55.

[208] Ronnett GV, Ramamurthy S, Kleman AM, Landree LE, Aja S. AMPK in the brain: its roles in energy balance and neuroprotection. *J Neurochem.* 2009;109 Suppl 1:17-23.

[209] Jibb LA, Richards JG. AMP-activated protein kinase activity during metabolic rate depression in the hypoxic goldfish, Carassius auratus. *J Exp Biol.* 2008;211(Pt 19):3111-22.

[210] Semenza GL, Wang GL. A nuclear factor induced by hypoxia via denovo protein-synthesis binds to the human erythropoietin gene enhancer at a site required for transcriptional activation. *Molecular and Cellular Biology*. 1992;12:5447-54.

[211] Wang GL, Semenza GL. Purification and characterization of hypoxia -inducible factor 1. *J Biol Chem.* 1995;270(3):1230-7.

[212] Purification and characterization of Hypoxia-inducible Factor 1, (1995).

[213] Semenza GL, Jiang B-H, Leung SW, Passantino R, Concordetţ J-P, Mairetţ P, et al., Hypoxia Response Elements in the Aldolase A, Enolase 1, and Lactate Dehydrogenase A Gene Promoters Contain Essential Binding Sites for Hypoxia-inducible Factor 1*. *The Journal of biological chemistry.* 1996;271:32529-37.

[214] Abnormal angiogenesis and responses to glucose and oxygen deprivation in mice lacking the protein ARNT., (1997).

[215] Ryan HE, Lo J, Johnson RS, Akazawa S, Unterman T, Metzger B, et al., HIF-1alpha is required for solid tumor formation and embryonic vascularization. *The EMBO Journal*. 1998;17:3005-15.

[216] Kline DD, Peng YJ, Manalo DJ, Semenza GL, Prabhakar NR. Defective carotid body function and impaired ventilatory responses to chronic hypoxia in mice partially deficient for hypoxia -inducible factor 1 alpha. *Proc Natl Acad Sci U S A*. 2002;99(2):821-6.

[217] Jiang BH, Rue E, Wang GL, Roe R, Semenza GL. Dimerization, DNA binding, and transactivation properties of hypoxia -inducible factor 1. *The Journal of biological chemistry*. 1996;271:17771-8.

[218] Wenger RH. Cellular adaptation to hypoxia: O_2-sensing protein hydroxylases, hypoxia-inducible transcription factors, and O_2-regulated gene expression. *The FASEB journal: official publication of the Federation of American Societies for Experimental Biology*. 2002;16:1151-62.

[219] Yuan G, Peng Y-J, Reddy VD, Makarenko VV, Nanduri J, Khan SA, et al., Mutual antagonism between hypoxia -inducible factor s 1α and 2α regulates oxygen sensing and cardio-respiratory homeostasis. *P Natl Acad Sci USA*. 2013;110:E1788-96.

[220] Epstein AC, Gleadle JM, McNeill LA, Hewitson KS, O'Rourke J, Mole DR, et al., C. elegans EGL-9 and mammalian homologs define a family of dioxygenases that regulate HIF by prolyl hydroxylation. *Cell*. 2001;107:43-54.

[221] Maxwell PH, Wiesener MS, Chang GW, Clifford SC, Vaux EC, Cockman ME, et al., The tumour suppressor protein VHL targets hypoxia -inducible factor s for oxygen -dependent proteolysis. *Nature*. 1999;399:271-5.

[222] Kornitzer D, Ciechanover a. Modes of regulation of ubiquitin -mediated protein degradation. *Journal of cellular physiology*. 2000;1 82:1-11.

[223] Ivan M, Kondo K, Yang H, Kim W, Valiando J, Ohh M, et al., HIFalpha targeted for VHL-mediated destruction by proline hydroxylation: implications for O2 sensing. *Science* (New York, NY). 2001;292:464-8.

[224] The VHL tumour-suppressor gene paradigm, (1998).

[225] Hewitson KS, McNeill LA, Riordan MV, Tian YM, Bullock AN, Welford RW, et al., Hypoxia-inducible factor (HIF) asparagine hydroxylase is identical to factor inhibiting HIF (FIH) and is related to the cupin structural family. *Journal of Biological Chemistry*. 2002;277:26351-5.

[226] McNeill LA, Hewitson KS, Gleadle JM, Horsfall LE, Oldham NJ, Maxwell PH, et al., The use of dioxygen by HIF prolyl hydroxylase (PHD1). *Bioorganic and Medicinal Chemistry Letters*. 2002;1 2:1 547-50.

[227] Hypoxia, hypoxia -inducible factor s (HIF), HIF hydroxylases and oxygen sensing, (2009).

[228] The von Hippel-Lindau protein, HIF hydroxylation, and oxygen sensing, (2005).

[229] Stroka DM, Burkhardt T, Desbaillets I, Wenger RH, Neil Da, Bauer C, et al., HIF-1 is expressed in normoxic tissue and displays an organ -specific regulation under systemic hypoxia. *The FASEB journal: official publication of the Federation of American Societies for Experimental Biology.* 2001;15:2445-53.

[230] Smith RW, Houlihan DF, Nilsson GE, Alexandre J. Tissue-specific changes in RNA synthesis in vivo during anoxia in crucian carp. *Am J Physiol.* 1999;277(3 Pt 2):R690-7.

[231] Smith RW, Houlihan DF, Nilsson GE, Brechin JG. Tissue-specific changes in protein synthesis rates in vivo during anoxia in crucian carp. *Am J Physiol.* 1996;271(4 Pt 2):R897-904.

[232] Wiesener MS, Jürgensen JS, Rosenberger C, Scholze CK, Hörstrup JH, Warnecke C, et al., Widespread hypoxia -inducible expression of HIF-2alpha in distinct cell populations of different organs. *The FASEB journal: official publication of the Federation of American Societies for Experimental Biology.* 2003;17:271-3.

[233] Chandel NS, Maltepe E, Goldwasser E, Mathieu CE, Simon MC, Schumacker PT. Mitochondrial reactive oxygen species trigger hypoxia -induced transcription. *Proc Natl Acad Sci U S A.* 1998;95(20):11715-20.

[234] Bell EL, Klimova TA, Eisenbart J, Schumacker PT, Chandel NS. Mitochondrial reactive oxygen species trigger hypoxia -inducible factor -dependent extension of the replicative life span during hypoxia. *Mol Cell Biol.* 2007;27(16):5737-45.

[235] Hochachka PW, Buck LT, Doll CJ, Land SC. Unifying theory of hypoxia tolerance: Molecular metabolic defense and rescue mechanisms for surviving oxygen lack. *P Natl Acad Sci USA.* 1996;93(18):9493-8.

[236] Semenza GL. Life with Oxygen. *Science.* 2007;318:62-4.

[237] Pamenter ME, Powell FL. Time Domains of the Hypoxic Ventilatory Response and Their Molecular Basis. *Compr Physiol.* 2016;6(3):1345-85.

[238] Pamenter ME, Powell FL. Signalling mechanisms of long term facilitation of breathing with intermittent hypoxia. *F1000prime reports*. 2013;5:23.

[239] Chung D, Dzal YA, Seow A, Milsom WK, Pamenter ME. Naked mole rats exhibit metabolic but not ventilatory plasticity following chronic sustained hypoxia. *Proc Biol Sci.* 2016;283(1827).

[240] Pamenter ME, Dzal YA, Milsom WK. Adenosine receptors mediate the hypoxic ventilatory response but not the hypoxic metabolic response in the naked mole rat during acute hypoxia. *Proc Biol Sci.* 2015;282(1800):20141722.

[241] Fregosi RF, Mitchell GS. Long-term facilitation of inspiratory intercostal nerve activity following carotid sinus nerve stimulation in cats. *J Physiol*. 1994;477 (Pt 3):469-79.

[242] Lee KZ, Fuller DD. Hypoxia-induced short-term potentiation of respiratory-modulated facial motor output in the rat. *Respir Physiol Neurobiol.* 2010;173(1):107-11.

[243] Eldridge FL, Millhorn DE, Eldridge FL, Millhorn DE. Oscillation, Gating, and Memory in the Respiratory Control System. *Comprehensive Physiology*. Hoboken, NJ, USA: John Wiley & Sons, Inc.; 2011.

[244] Bisgard GE. Carotid body mechanisms in acclimatization to hypoxia. *Respir Physiol.* 2000;121(2-3):237-46.

[245] Dwinell MR, Huey KA, Powell FL. Chronic hypoxia induces changes in the central nervous system processing of arterial chemoreceptor input. *Adv Exp Med Biol.* 2000;475:477-84.

[246] Barnard P, Andronikou S, Pokorski M, Smatresk N, Mokashi A, Lahiri S. Time-dependent effect of hypoxia on carotid body chemosensory function. *J Appl Physiol.* 1987;63(2):685-91.

[247] Vizek M, Pickett CK, Weil JV. Increased carotid body hypoxic sensitivity during acclimatization to hypobaric hypoxia. *J Appl Physiol.* 1987;63(6):2403-10.

[248] Chen J, He L, Liu X, Dinger B, Stensaas L, Fidone S. Effect of the endothelin receptor antagonist bosentan on chronic hypoxia -induced morphological and physiological changes in rat carotid body. *Am J Physiol Lung Cell Mol Physiol*. 2007;292(5):L1257-62.

[249] Forster HV, Bisgard GE, Klein JP. Effect of peripheral chemoreceptor denervation on acclimatization of goats during hypoxia. *J Appl Physiol*. 1981;50(2):392-8.

[250] Smith CA, Bisgard GE, Nielsen AM, Daristotle L, Kressin NA, Forster HV, et al., Carotid bodies are required for ventilatory acclimatization to chronic hypoxia. *J Appl Physiol*. 1986;60(3):1003-10.

[251] Roeggla G, Roeggla M, Wagner A, Laggner AN. Poor ventilatory response to mild hypoxia may inhibit acclimatization at moderate altitude in elderly patients after carotid surgery. *British journal of sports medicine*. 1995;29:110-2.

[252] Tatsumi K, Pickett CK, Weil JV. Attenuated carotid body hypoxic sensitivity after prolonged hypoxic exposure. *J Appl Physiol*. 1991;70(2):748-55.

[253] Weil JV. Ventilatory Control at High Altitude. *Handbook of Physiology, The Respiratory System, Control of Breathing*. 2011.

[254] Pamenter ME, Carr JA, Go A, Fu Z, Reid SG, Powell FL. Glutamate receptors in the nucleus tractus solitarius contribute to ventilatory acclimatization to hypoxia in rat. *J Physiol*. 2014;592(8):1839-56.

[255] Pamenter ME, Go A, Fu Z, Powell FL. No evidence of a role for neuronal nitric oxide synthase in the nucleus tractus solitarius in ventilatory responses to acute or chronic hypoxia in awake rats. *J Appl Physiol* (1985). 2015;118(6):750-9.

[256] Pamenter ME, Nguyen J, Carr JA, Powell FL. The effect of combined glutamate receptor blockade in the NTS on the hypoxic ventilatory response in awake rats differs from the effect of individual glutamate receptor blockade. *Physiological reports*. 2014;2(8).

[257] Chung D, Dzal YA, Seow A, Milsom WK, Pamenter ME. Naked mole rats exhibit metabolic but not ventilatory plasticity following chronic sustained hypoxia. *Proc Biol Sci.* 2016;283(1827):20160216.

[258] Hempleman SC. Sodium and potassium current in neonatal rat carotid body cells following chronic in vivo hypoxia. *Brain Res.* 1995;699(1):42-50.

[259] Hempleman SC. Increased calcium current in carotid body glomus cells following in vivo acclimatization to chronic hypoxia. *J Neurophysiol.* 1996;76(3):1880-6.

[260] Pascual O, Denavit-Saubie M, Dumas S, Kietzmann T, Ghilini G, Mallet J, et al., Selective cardiorespiratory and catecholaminergic areas express the hypoxia -inducible factor -1alpha (HIF-1alpha) under in vivo hypoxia in rat brainstem. *Eur J Neurosci.* 2001;14(12):1981-91.

[261] Helton R, Cui J, Scheel JR, Ellison Ja, Ames C, Gibson C, et al., Brain-specific knock-out of hypoxia -inducible factor -1alpha reduces rather than increases hypoxic-ischemic damage. *The Journal of neuroscience: the official journal of the Society for Neuroscience.* 2005;25:4099-107.

[262] Bavis RW, Powell FL, Bradford A, Hsia CC, Peltonen JE, Soliz J, et al., Respiratory plasticity in response to changes in oxygen supply and demand. *Integr Comp Biol.* 2007;47(4):532-51.

[263] Peng YJ, Nanduri J, Khan SA, Yuan G, Wang N, Kinsman B, et al., Hypoxia-inducible factor 2alpha (HIF-2alpha) heterozygous-null mice exhibit exaggerated carotid body sensitivity to hypoxia, breathing instability, and hypertension. *Proc Natl Acad Sci U S A.* 2011;108(7):3065-70.

[264] Yuan G, Khan SA, Luo W, Nanduri J, Semenza GL, Prabhakar NR. Hypoxia-inducible factor 1 mediates increased expression of NADPH oxidase-2 in response to intermittent hypoxia. *J Cell Physiol.* 2011;226(11):2925-33.

[265] Yuan G, Nanduri J, Khan S, Semenza GL, Prabhakar NR. Induction of HIF-1?? expression by intermittent hypoxia: Involvement of

NADPH oxidase, Ca2+ signaling, prolyl hydroxylases, and mTOR. *Journal of Cellular Physiology*. 2008;217:674-85.

[266] Dhillon DP, Barer GR, Walsh M. The enlarged carotid body of the chronically hypoxic and chronically hypoxic and hypercapnic rat: a morphometric analysis. *Q J Exp Physiol*. 1984;69(2):301-17.

[267] McGregor KH, Gil J, Lahiri S. A morphometric study of the carotid body in chronically hypoxic rats. *J Appl Physiol*. 1984;57(5):1430-8.

[268] Chen J, He L, Dinger B, Fidone S. Cellular mechanisms involved in rabbit carotid body excitation elicited by endothelin peptides. *Resp Physiol*. 2000;121:13-23.

[269] Chen J, He L, Dinger B, Stensaas L, Fidone S. Role of endothelin and endothelin A-type receptor in adaptation of the carotid body to chronic hypoxia. *Am J Physiol Lung Cell Mol Physiol*. 2002;282(6):L1314-23.

[270] Pawar A, Nanduri J, Yuan G, Khan Sa, Wang N, Kumar GK, et al., Reactive oxygen species -dependent endothelin signaling is required for augmented hypoxic sensory response of the neonatal carotid body by intermittent hypoxia. *American journal of physiology Regulatory, integrative and comparative physiology*. 2009;296:R735-R42.

[271] Chen J, He L, Dinger B, Stensaas L, Fidone S. Chronic hypoxia upregulates connexin43 expression in rat carotid body and petrosal ganglion. *J Appl Physiol*. 2002;92(4):1480-6.

[272] Kuwaki T, Cao WH, Kurihara Y, Kurihara H, Ling GY, Onodera M, et al., Impaired ventilatory responses to hypoxia and hypercapnia in mutant mice deficient in endothelin-1. *The American journal of physiology*. 1996;270:R1279-86.

[273] Neubauer JA, Sunderram J. Heme oxygenase-1 and chronic hypoxia. *Respir Physiol Neurobiol*. 2012;184(2):178-85.

[274] Sunderram J, Semmlow J, Thakker-Varia S, Bhaumik M, Hoang-Le O, Neubauer JA. Heme oxygenase-1-dependent central cardiorespiratory adaptations to chronic hypoxia in mice. *Am J Physiol Regul Integr Comp Physiol*. 2009;297(2):R300-12.

[275] Jyung RW, LeClair EE, Bernat RA, Kang TS, Ung F, McKenna MJ, et al., Expression of angiogenic growth factor s in paragangliomas. *The Laryngoscope.* 2000;110:161-7.

[276] Fung M-L, Tipoe GL. Role of HIF-1 in physiological adaptation of the carotid body during chronic hypoxia. *Advances in experimental medicine and biology.* 2003;536:593-601.

[277] Ward ME, Toporsian M, Scott JA, Teoh H, Govindaraju V, Quan A, et al., Hypoxia induces a functionally significant and translationally efficient neuronal NO synthase mRNA variant. *Journal of Clinical Investigation.* 2005;115:3128-39.

[278] Carotid body mechanisms in acclimatization to hypoxia, (2000).

[279] El Hasnaoui-Saadani R, Alayza RC, Launay T, Pichon A, Quidu P, Beaudry M, et al., Brain stem NO modulates ventilatory acclimatization to hypoxia in mice. *J Appl Physiol.* 2007;103(5):1506-12.

[280] Post JM, Hume JR, Archer SL, Weir EK. Direct role for potassium channel inhibition in hypoxic pulmonary vasoconstriction. *Am J Physiol.* 1992;262:C882-C90.

[281] Sweeney M, Yuan JX. Hypoxic pulmonary vasoconstriction: role of voltage-gated potassium channels. *Respiratory research.* 2000;1:40-8.

[282] Fandrey J. Oxygen-dependent and tissue -specific regulation of erythropoietin gene expression. *American journal of physiology Regulatory, integrative and comparative physiology.* 2004;286:R977-88.

[283] Lee FS. Genetic causes of erythrocytosis and the oxygen -sensing pathway. *Blood Reviews.* 2008;22:321-32.

[284] Ang SO, Chen H, Hirota K, Gordeuk VR, Jelinek J, Guan Y, et al., Disruption of oxygen homeostasis underlies congenital Chuvash polycythemia. *Nature genetics.* 2002;32:614-21.

[285] Smith TG, Brooks JT, Balanos GM, Lappin TR, Layton DM, Leedham DL, et al., Mutation of von Hippel-Lindau tumour suppressor and human cardiopulmonary physiology. *PLoS Med.* 20 06;3(7):e290.

[286] Pugh CW, Ratcliffe PJ. Regulation of angiogenesis by hypoxia: role of the HIF system. *Nature medicine.* 2003;9:677-84.

[287] Lok CN, Ponka P. Identification of a hypoxia response element in the transferrin receptor gene. *Journal of Biological Chemistry.* 1999;274:24147-52.

[288] Szpir M. Why did the crow cross the road? *Am Sci.* 2003;91(3):215-6.

[289] Buck LT, Hochachka PW, Schon A, Gnaiger E. Microcalorimetric measurement of reversible metabolic suppression induced by anoxia in isolated hepatocytes. *Am J Physiol.* 1993;265(5 Pt 2):R1014-9.

[290] Buck LT, Land SC, Hochachka PW. Anoxia-tolerant hepatocytes: model system for study of reversible metabolic suppression. *Am J Physiol Regul Integr Comp Physiol.* 1993;265(R49-R56).

[291] Pamenter ME, Buck LT. delta-Opioid receptor antagonism induces NMDA receptor-dependent excitotoxicity in anoxic turtle cortex. *J Exp Biol.* 2008;211(Pt 21):3512-7.

[292] Pamenter ME, Hogg DW, Buck LT. Endogenous reductions in N-methyl-d-aspartate receptor activity inhibit nitric oxide production in the anoxic freshwater turtle cortex. *FEBS Lett.* 2008;582:1738-42.

[293] Pamenter ME, Shin DS, Buck LT. Adenosine A1 receptor activation mediates NMDA receptor activity in a pertussis toxin -sensitive manner during normoxia but not anoxia in turtle cortical neurons. *Brain Res.* 2008;1213:27-34.

[294] Pamenter ME, Shin DS, Buck LT. AMPA receptors undergo channel arrest in the anoxic turtle cortex. *Am J Physiol Regul Integr Comp Physiol.* 2008;294(2):R606-13.

[295] Pamenter ME, Shin DS, Cooray M, Buck LT. Mitochondrial ATP - sensitive K+ channels regulate NMDAR activity in the cortex of the anoxic western painted turtle. *J Physiol.* 2008;586(4):1043-58.

[296] Pamenter ME. Delta -opioid receptors and glutamatergic toxicity. In: Xia Y, editor. *Neural function of the delta-opioid receptor.* New York: Springer; 2014.

[297] Shin DS, Wilkie MP, Pamenter ME, Buck LT. Calcium and protein phosphatase 1/2A attenuate N-methyl-D-aspartate receptor activity in

the anoxic turtle cortex. *Comp Biochem Physiol A Mol Integr Physiol.* 2005;142(1):50-7.

[298] Wilkie MP, Pamenter ME, Alkabie S, Carapic D, Shin DS, Buck LT. Evidence of anoxia -induced channel arrest in the brain of the goldfish (Carassius auratus). *Comp Biochem Physiol C Toxicol Pharmacol.* 2008.

[299] Buck LT, Hogg DW, Rodgers-Garlick C, Pamenter ME. Oxygen sensitive synaptic neurotransmission in anoxia -tolerant turtle cerebrocortex. In: Gotsiridze-Columbus N, editor. *Arterial Chemoreception: From Molecules to Systems*. New York: Springer; 2 011.

[300] Pamenter ME, Dzal YA, Milsom WK. Profound metabolic depression in the hypoxia -tolernat naked mole rat. *FASEB J.* 2014;28(1):879.2.

[301] Peterson BL, Larson J, Buffenstein R, Park TJ, Fall CP. Blunted neuronal calcium response to hypoxia in naked mole -rat hippocampus. *PLoS One*. 2012;7(2):e31568.

[302] Peterson BL, Park TJ, Larson J. Adult naked mole -rat brain retains the NMDA receptor subunit GluN2D associated with hypoxia tolerance in neonatal mammals. *Neurosci Lett.* 2012;506(2):342-5.

[303] Bryan JD, Hill LG, Neill WH. Interdependence of acute temperature preference and respiration in the plains minnow. *Transactions of the American Fisheries Society*. 1984;113:557-62.

[304] Schurmann H, Steffensen JF, Lomholt JP. The influence of hypoxia on the preferred temperature of rainbow trout Oncorhynchus mykiss. *Journal of Experimental Biology*. 1991;157:75-86.

[305] Hicks JW, Wood SC. Temperature regulation in lizards: effects of hypoxia. *The American journal of physiology*. 1985;248:R595-600.

[306] Dupré RK, Wood SC. Behavioral temperature regulation by aquatic ectotherms durIng hypoxia. *Can J Zool*. 1988;66:2649-52.

[307] Mortola JP, Feher C. Hypoxia inhibits cold-induced huddling in rat pups. *Resp Physiol*. 1998;113(3):213-22.

[308] Wood SC. Interactions between Hypoxia and Hypothermia. *Annual Review of Physiology*. 1991;53:71-85.

[309] Skovgaard N, Hicks JW, Wang T. Oxygen uptake and transport in air breathers. In: Nilsson GE, editor. *Respiratory Physiology of Vertebrates.* Cambridge: Cambridge University Press; 2010. p. 95-128.

[310] Sowter HM, Ratcliffe PJ, Watson P, Greenberg AH, Harris AL. HIF-1-dependent regulation of hypoxic induction of the cell death factors BNIP3 and NIX in human tumors. *Cancer Research.* 2001;61:6669-73.

[311] Papandreou I, Cairns RA, Fontana L, Lim AL, Denko NC. HIF-1 mediates adaptation to hypoxia by actively downregulating mitochondrial oxygen consumption. *Cell Metabolism.* 2006;3:187-97.

In: Chemosensory Sensors and Systems
Editor: Maximilian Brandt
ISBN: 978-1-53612-406-4

Chapter 2

CHEMICAL SENSOR DEVELOPMENT BY SRO.CNT NANOCOMPOSITES

Mohammed Muzibur Rahman, PhD
Mohammad M. Hussain and Abdullah M. Asiri, PhD
Chemistry Department and Center of Excellence
for Advanced Material Research, Faculty of Sciences,
King Abdulaziz University, Saudi Arabia

ABSTRACT

Nanocomposites of strontium oxide nanoparticles decorated on carbon nanotube (SrO.CNT NCs) were prepared in basic medium using a wet-chemical method. The SrO.CNT NCs were investigated using Fourier transform infra-red spectroscopy (FTIR), Ultraviolet- visible spectroscopy (UV-Vis.); Field emission scanning electron microscopy (FESEM) equipped Electron dispersive spectroscopy (EDS), Energy dispersive X-ray spectroscopy (XPS), and X-ray powder diffraction (XRD) techniques. A selective hydrazine sensor onto a flat glassy carbon electrode (GCE) was fabricated with a thin-layer of NCs. Improved electrochemical performances such as higher sensitivity, limit of detection; linear dynamic range, response time, and long-term stability towards hydrazine were achieved using the modified chemical sensor.

The calibration curve was found linear over a wide range of hydrazine concentration. The sensitivity and detection limit were calculated based on slope/surface area of GCE, and signal to noise ratio ~ 3 respectively. It is the first time report regarding hydrazine detection using current *vs* voltage (I-V) process by means of SrO.CNT NCs modified GCE. The synthesis of SrO.CNT NCs using a wet-chemical procedure is a good way of establishing chemical sensor based on nano-materials (NMs) for toxic elements. It is also commenced a potential development of hydrazine sensor using SrO.CNT NCs by a simplistic and dependable I-V technique for the prospective appliance of harmful agents detection in environmental and health care area. Finally, the effectiveness and efficiency of the proposed selective chemical sensors could utilized significantly and effectively for the detection of toxic hazardous chemicals for safety green environment and healthcare fields in broad scales.

INTRODUCTION

Chemical sensor is a self-contained analytical device that can provide information about the chemical composition of its environment, that is, a liquid or gas phase. The information is provided in the form of a measurable physical signal that is correlated with the concentration of a certain chemical species (termed as analyte). Two main steps are involved in the functioning of a chemical sensor, namely, recognition and transduction. In the recognition step, analyte molecules interact selectively with receptor molecules or sites included in the structure of the recognition element of the sensor. Consequently, a characteristic physical parameter varies and this variation is reported by means of an integrated transducer that generates the output signal. A chemical sensor based on recognition material of biological nature is a biosensor. However, as synthetic bio-mimetic materials are going to substitute to some extent recognition biomaterials, a sharp distinction between a biosensor and a standard chemical sensor is superfluous. Typical bio-mimetic materials used in sensor development are molecularly imprinted polymer sandaptamers [1-3].

Carbon material is most notably used to reinforce composite materials, particularly the class of materials known as carbon fiber or graphite reinforced polymers. Non-polymer materials can also be used as the matrix for carbon fibers. Due to the formation of metal carbide sand corrosion considerations, carbon has seen limited success in metal matrix composite applications. Reinforced carbon-carbon consists of carbon fiber-reinforced graphite, and is used structurally in high-temperature applications. The fiber also finds use in filtration of high-temperature gases, as an electrode with high surface area and impeccable corrosion resistance, and as an anti-static component. Molding a thin layer of carbon fibers significantly improves fire resistance of polymers or thermoset composites because a dense, compact layer of carbon fibers efficiently reflects heat. The increasing use of carbon fiber composites is displacing aluminum from aerospace applications in favor of other metals because of galvanic corrosion issues [4-6].

There is significant interest in developing electrochemical techniques for determination of neuro-transmitters such as hydrazine, phenolic compounds, penicillin, dopamine etc. Biochemical is ever-present neurotransmitter in mammalian brain tissues that plays an important physiological role in the functioning of central nervous, renal, hormonal and cardiovascular system s as an extra cellular chemical messenger [7-9]. Dopamine (DA) belongs to the family of inhibitory neurotransmitters; its function is to regulate neural interactions by reducing the permeability of gap junctions between adjacent neurons of the same type. In addition, neuro-degeneration of DA-containing neurons contributes to late-onset neurological disease s, including Parkinson's and Alzheimer's diseases, and possibly to normal ageing of the brain [10]. Besides, numerous reports have shown its coexistence in biological systems [11-13]. Therefore, investigation of neurological behavior and also simultaneous determination of DA is of great importance for the elucidation of its precise physiological functions. A range of analytical techniques such as chromatographic methods, mass spectroscopy [14, 15], spectrophotometry [16, 17], and chemiluminesence [18, 19] are reported in the literature for detection of DA. However these methods suffer from some disadvantages including long analysis times, high costs, the requirement for sample pretreatment, and in some cases low sensitivity and selectivity. These disadvantages probably make them unsuitable for routine analysis. The advantages of electrochemical methods for determination of DA include low cost, high sensitivity and short measurement time [20, 21]. However, the electrochemical measurement of neurotransmitter concentrations has been mainly unsatisfactory due to the inability of the electrodes employed to separate the potentials of these species sufficiently to allow for accurate detection. Ascorbic acid (AA) is usually present in vivo at concentrations 100-1000 times higher than the neurotransmitters. Obviously, it is necessary to develop selective and sensitive techniques to resolve these problems. Therefore, various electrochemical approaches have been made to overcome these difficulties for the determination of DA [22, 23]. The attractive features of the multi-walled carbon nano-tubes (MWCNTs) including their unique mechanical and electrical properties are significant advantages in the design and development of the electrochemical sensors [24, 25]. Furthermore, MWCNTs modified by the adsorption of small organic molecules or metal nanoparticles via redox or electro-deposition can be used to determine H_2O_2 iodate dopamine, organic small molecules, inorganic ion and even widely used as catalytic electrodes of fuel cell s [26-28]. Multiwalled carbon nanotubes (MWCNT) modified glassy carbon electrodes GCEs exhibited signals enhanced by about five fold in the detection of dopamine in the presence of AA [29, 30] reviewed the use of MWCNTs to obtain enhanced signals in the detection of substances such as carbohydrates, nucleic acid s, glucose, pesticides, and serotonin, with similar reports relating to trace metals [31, 32] and nitroaromatic compounds [33]. There are several literatures on the electro-oxidation of dopamine on modified electrodes [35-38].

Nanotechnology can enable sensors to detect very small amounts of chemical vapours. Various types of detecting elements, such as carbon nanotubes, zinc oxide nanowires or palladium nanoparticles can be used in nanotechnology -based sensors. These detecting elements change their electrical characteristics, such as resistance or capacitance, when they absorb a gas molecule. The chemical sensor plays an essential role in the fields of environmental conservation and monitoring, disaster and disease prevention, and industrial analysis. A typical chemical sensor is a device that transforms chemical information in a selective and reversible way, ranging from the concentration of a specific sample component to total composition analysis, into an analytically useful signal. Much research work has been performed to achieve a chemical sensor with such excellent qualities as quick response, low cost, small size, superior sensitivity, good reversibility and selectivity, and excellent detection limit [39, 40]. This collection of up-to-date information and the latest research progress on chemical sensor will provide valuable references and learning materials for all those working in the field of chemical sensors. In the wide portfolio of sensors manufactured are chemical gas sensor solutions for appliance, automotive, building technology, consumer and industrial applications. With a track record going back almost two decades, it is a world-leading supplier of safety, energy efficiency and comfort solutions for global mass markets. Because of the small size of nanotubes, nanowires, or nanoparticles, a few gas molecules are sufficient to change the electrical properties of the sensing elements. This allows the detection of a very low concentration of chemical vapors [41]. The goal is to have small, inexpensive sensors that can sniff out chemicals just as dogs are used in airports to smell the vapors given off by explosives or drugs. The capability of producing small, inexpensive sensors that can quickly identify a chemical vapor provides a kind of nano-bloodhound that doesn't need sleep or exercise which can be useful in a number of ways. An obvious application is to mount these sensors throughout an airport, or any facility with security concerns, to check for vapors given off by explosive devices. Researchers were developed explosive sensors using carbon nanotubes, where the bond with polymers to the carbon nanotubes to tune the nanotubes to different explosives or chemical vapours. They also have developed a sensor using carbon nanotubes embedded in a gel; that can be injected under the skin to monitor the level of nitric oxide in the bloodstream. The level of nitric oxide is important because it indicates inflammation, allowing easy monitoring of inflammatory disease s. In tests with laboratory mice the sensor remained functional for over a year. It was demonstrated a method of spraying carbon nanotubes onto flexible plastic surfaces to produce sensors [42, 43].

LITERATURE SURVEY

Chemical pollutant, hydrazine is one of the most generally used compounds in the industrial and environmental samples due to its toxic effect s in environment and human beings. It is extensively utilized as antioxidants, rocket fuels, catalysts, emulsifier, corrosion inhibitor, reducing agents, pesticides, and plant growth regulators [44, 45]. It is a neurotoxin, and hence produces carcinogenic and mutagenic effects causing damage to lungs, liver, and kidneys, respiratory tract infection and long-term effects on the central nervous system [46]. In pharmaceuticals also, it has much significance as it has been acknowledged as carcinogenic and hepatotoxic chemicals which affects the liver and brain [47, 48]. Electro-oxidation of hydrazine is the basis of an established fuel cell, owing to its high capacity and no contamination. In addition to this, hydrazine is widely used as high-energy propellants in rockets and spacecrafts by military and aerospace industries [49]. Owing to aforementioned applications in industry, environment and pharmacology, it is very significant to fabricate a reliable, cheap and effective technique for the efficient detection of hydrazine. Among several detection methodologies, the electrochemical I-V method offers an opportunity for portable, cheap and rapid methodologies. Therefore, numerous chemically modified electrodes, based on different nanostructure materials, semiconductor doped or un-doped nanomaterials, transition metal oxides, electro-catalytic moieties (electron -mediator species), has thus been developed for the detection of hydrazine and reported in the literature [50, 51]. Recently, scientists have demonstrated the use of semiconductor nanostructures as electron mediators to modify the electrodes for the electrochemical detection of toxic hydrazine [52, 53].

The safety is a major concern for environment and health & is a vital issue to study in perspective of using chemical sensors for the identification of poisonous materials through a well-established method. Due to the high active surface area and spherical size to volume ratio in comparison with conventional nanomaterials at the micro to nano ranges, nanostructure materials semiconductor is very efficient and sensitive. The nanostructure of metal oxides having nice properties, fabrication of chemical sensors, high active surface area, high porosity, permeability, quantum confinement effect and stability is a great deal of attention at present [54, 55]. Having the properties of high response, large surface area, lower charge, measoporous nature and portability, sensor -based metal oxide conjugated carbon nanotube composites are widely employed for the monitoring of chemical process, air/water contamination and toxic agents in the environment [56, 57]. Separation of hazardous materials from industrial waste water is one of the vital issues in the environmental and biological science arena. Different techniques were reported for the removing of carcinogenic molecules from the industrial waste water. But some issues are still remaining unsettled, efficiently removing of hazardous agents and re-usability of the NCs materials including preparation of the green NCs at facile and low cost. The mesoporous nature of the NCs material allows facile recycling without major deterioration of sensor efficiency and potentiality. A good absorption/adsorption power of the hybrid NCs along with other merits such as easy separation, environmental friendly composition and re-usability makes it a suitable sensor for the removing of target toxic materials from industrial as well as environmental wastes [58, 59].

Hydrazine is a colorless inorganic compound having antioxidant and reducing properties. It has wide applications in various fields such as aircraft fuels, antioxidants, blowing agent s, catalyst, corrosion inhibition, emulsifiers, explosives, fuel cell s, heat stabilizers, insecticides, metal films production, oxygen scavengers, pharmaceutical intermediates, photographic developer, pesticides, plastic polymers, plant growth regulators, plutonium extraction, rocket propellant and textile dyes [60, 61]. Beyond of these applications, hydrazine is a toxic agent responsible for blood abnormalities, carcinogen, cell mutagenicity, coma, dizziness, hepatoxicity, nausea, neurotoxic activity, pulmonary edema, temporary blindness including damage of various organs such as brain, DNA, eye, kidney, liver, lungs, nose, nervous and respiratory system and throat infection [62-68]. Different conventional analytical methods are used for the determination of hydrazine including amperometry, chromatography (Gas and HPLC), capillary electrophoresis, chemiluminescence, colorimetry, coulometry, electrochemical technique, fluorescence, mass spectroscopy, potentiometry, spectrophotometry and titrimetry [69-71]. But most of the techniques are complex for on-site analysis, costly and time consuming. So, in perspective of environmental and biological importance, it is necessary to develop selective and sensitive analytical technique for the identification and determination of hydrazine. Detection of hydrazine using fabricated electrode with NCs is easy to implement along with high sensitivity, low cost and real time analysis [72].

An oxide is a chemical compound that contains at least one oxygen atom and one other element in its chemical formula. Metal oxides typically contain an ion-oxygen in the oxidation state of -2. Most of the Earth's crust consists of solid oxides, the result of elements being oxidized by the oxygen in air or in water. Hydrocarbon combustion affords the two principal carbon oxides, carbon monoxide and carbon dioxide. Even materials considered pure elements often develop an oxide coating [73, 74]. For example, aluminum foil develops a thin skin of MOx that protects the foil from further corrosion. Different oxides of the same element are distinguished by Roman numerals denoting their oxidation number, e.g., iron (II) oxide versus iron(III) oxides. Due to its electro-negativity, oxygen forms stable chemical bonds with almost all elements to give the corresponding oxides [75, 76].Noble metals (such as gold or platinum) are prized because they resist direct chemical combination with oxygen, and substances like gold(III) oxide must be generated by indirect routes [77].

Nano composite is a multiphase solid material where one of the phases has one, two or three dimensions of less than 100 nm, or structures having nano-scale repeat distances between the different phases that make up the material. In the broadest sense this definition can include porous media, colloids, copolymers, but is more usually taken to mean the solid combination of a bulk matrix and nano-dimensional phase(s) differing in properties due to dissimilarities in structure and chemistry. The mechanical, electrical, thermal, optical, electrochemical, catalytic properties of the nano-composite will differ markedly from that of the component materials. Size limits for these effects have been proposed <5 nm for catalytic activity [78, 79], <20 nm for making a hard magnetic material soft, <50 nm for refractive index changes, and <100 nm for achieving super paramagnetism, mechanical strengthening or restricting matrix dislocation movement. Nano-composites are found in nature, for example in the structure of theabalone shelland bone [80, 81]. The use of nanoparticle rich materials long predates the understanding of the physical and chemical nature of these materials [82, 83], investigated the origin of the depth of color and the resistance to acids and bio-corrosion of Maya blue paint, attributing it to a nanoparticle mechanism. From the mid-1950s nanoscale organo-clays have been used to control flow of polymer solutions (e.g., as paint viscosifiers) or the constitution of gels (e.g., as a thickening substance in cosmetics, keeping the preparations in homogeneous form). By the 1970s polymer/clay composites were the topic of textbooks [84, 85], although the term "nano-composites" was not in common use. In mechanical terms, nano-composites differ from conventional composite materials due to the exceptionally high surface to volume ratio of the reinforcing phase and/or its exceptionally high aspect ratio. The reinforcing material can be made up of particles (e.g., minerals), sheets (e.g., exfoliated clay stacks) or fibres (e.g., carbon nanotubes or electro-spun fibers). The area of the interface between the matrix and reinforcement phase(s) is typically an order of magnitude greater than for conventional composite materials. The matrix material properties are significantly affected in the vicinity of the reinforcement [86, 87],note that with polymer nano-composites, properties related to local chemistry, degree of thermoset cure, polymer chain mobility, polymer chain conformation, degree of polymer chain ordering or crystallinity can all vary significantly and continuously from the interface with the reinforcement into the bulk of the matrix.

In previous studies SrO was used as coating materials on carbon nanotubes [88], additive coloration [89, 90], doping agent [91, 92], Aldol addition, kinetic, paramagnetic resonance, IR, Mass, and UV-Vis spectroscopic study [93-95], cement forming ion [96, 97], heterogeneous catalyst [98], isomerization effect moiety [99], preparation of thin films [100], deposition of magnetron sputter [101], photocatalytic activity [102, 103]. Here, hydrazine is extremely toxic and usually serious to health and environment, it is immediately required the detection by using a reliable method with SrO.CNT NCs using GCE electrode. The investigation of hydrazine by thin SrO.CNT NCs deposited thin films on GCE will be prepared and studied in details of the chemical sensors. The easy-coating method for the construction of SrO.CNT NCs thin-layer within conducting binding-agents will be executed for preparation of films onto GCE. In this approach, mesoporous SrO.CNT NCs fabricated films with conducting binders were utilized towards the target carcinogenic analytes using reliable I-V method. It is confirmed that the fabricated hydrazine sensor is unique and noble research work for ultra-sensitive recognition with active SrO.CNT NCs onto GCE in short response -time. In this work, synthesis, electrochemical characterization, screening for determination of hydrazine using fabricated SrO.CNT NCs has been reported.

METHODOLOGY

The used chemicals of the analytical grade, acetone, 4-aminophenol, ammonium hydroxide, benzaldehyde, bisphenol A, carbon nanotubes, disodium phosphate, ethanol, hydrazine, melamine, methanol, 4-methoxyphenol, monosodium phosphate, nafion (5% ethanolic solution), 2-nitrophenol, sodium hydroxide and strontium chloride were purchased from Sigma-Aldrich Company. FTIR and UV-Vis spectra of the dried

SrO.CNT NCs were recorded on a Thermo scientific NICOLET iS50 FTIR spectrometer (Madision, WI, USA) and 300 UV/Visible spectrophotometer (Thermo scientific) respectively. The XPS measurements were carried out on a K-α spectrometer (Thermo scientific, K-α 1066) with an excitation radiation source (A1Kα, Beam spot size = 300.0 μm, pass energy = 200.0 eV, pressure ~ 10^{-8} torr) for the calculation of binding energies (eV) among Sr, C and O. The arrangement, morphology and particle size of the SrO.CNT NCs were investigated using field emission scanning electron microscopy (FESEM, JEOL, JSM-7600F, Japan). X-ray diffraction (XRD) examination was performed under ambient conditions to recognize the crystalline pattern of SrO.CNT NCs. I-V was conducted to detect hydrazine at a selective point using Keithley electrometer (6517A, USA) at normal condition by fabricated SrO.CNT NCs. A current–voltage characteristic or I–V curve (current–voltage curve) is a relationship, typically represented as a chart or graph, between the electric current through a circuit, device, or material, and the corresponding voltage, or potential difference across it. In electronics, the relationship between the direct current (DC) through an electronic device and the DC-voltage across its terminals is called a current–voltage characteristic of the device. Electronic engineers use these charts to determine basic parameters of a device and to model its behavior in an electrical circuit. These characteristics are also known as I-V curves, referring to the standard symbols for current and voltage. In electronic components with more than two terminals, such as vacuum tubes and transistors, the current-voltage relationship at one pair of terminals may depend on the current or voltage on a third terminal. This is usually displayed on a more complex current–voltage graph with multiple curves, each one representing the current-voltage relationship at a different value of current or voltage on the third terminal [104, 105]. The simplest IV characteristic involves a resistor, which according to Ohm's Law exhibits a linear relationship between the applied voltage and the resulting electric current. However, even in this case environmental factors such as temperature or material characteristics of the resistor can produce a non-linear curve [106, 107].

Preparation of Nanocomposites

A buffer solution (more precisely, pH buffer or hydrogen ion buffer) is an aqueous solution consisting of a mixture of a weak acid and its conjugate base, or vice versa. Its pH changes very little when a small or moderate amount of strong acid or base is added to it and thus it is used to prevent changes in the pH of a solution. Buffer solutions are used as a means of keeping pH at a nearly constant value in a wide variety of chemical applications. Many life forms thrive only in a relatively small pH range so they utilize a buffer solution to maintain a constant pH. In nature, the bicarbonate buffering system is used to regulate the pH of blood. Phosphate-buffered saline (abbreviated PBS) is a buffer solution commonly used in biological research. It is a water -based salt solution containing sodium hydrogen phosphate, sodium chloride and, in some formulations, potassium chloride and potassium di-hydrogen phosphate. The osmolarity and ion concentrations of the solutions match those of the human body (isotonic). The wet-chemical procedure [97, 108] is a primary stage for the preparation of materials and powders using a liquid phase which is a typical solid-state synthesis technique and widely used in the synthesis of doped or un-doped NMs. The products will achieve smaller grains along with shorter duration of phase formation. Strontium chloride ($SrCl_2.6H_2O$), carbon nanotube (CNT) and sodium hydroxide (NaOH) were used in the preparation of SrO.CNT NCs as active reacting agents. Based on this technique, CNT (1.0 wt %, 0.25 μg) was added under continuous stirring with dissolved $SrCl_2.6H_2O$ (0.1 M, 2.67 g) in distilled water (100 mL) at an erlenmeyer flask (250.0 mL); the pH of the resultant mixture was adjusted to 10.13 with addition of NaOH and then placed for continuous stirring at 90.0 °C. The flask washed thoroughly with water and acetone subsequently after continuous stirring (6 h) and kept for solvent evaporation at room temperature. The resultant gray products (SrO.CNT NCs) were dried in the oven at 60.0°C (24 h), and then grinded into powders. The powdered SrO.CNT NCs was again dried in the oven at 60°C for 22 h and afterward used for characterizations such as elemental, morphological, optical and structural, and application as a chemical sensor.

Strontium oxide nanoparticles (SrO NPs) without CNT were also prepared using the same procedure. The NCs growth mechanism is presented in Scheme 1. In SrO.CNT NC growth method, initially SrO nucleus growth takes place by itself and mutual-aggregation, nano-crystal re-aggregates and formed aggregated SrO nanocrystal using Ostwald-ripening method. Nanocrystal crystallizes and re-aggregates with each other counter parts in presence of disperse CNT through "Vander Waal's forces" and reforms SrO.CNT porous morphology [109]. The possible reaction mechanism s for the synthesis of SrO.CNT NCs are shown as follows.

$$NaOH_{(s)} \rightarrow Na^{+}{}_{(aq)} + OH^{-}{}_{(aq)} \qquad (i)$$

$$SrCl_2 \rightarrow Sr^{2+}{}_{(aq)} + 2Cl^{-}{}_{(aq)} \qquad (ii)$$

$$2Na^{+}{}_{(aq)} + 2OH^{-}{}_{(aq)} + Sr^{2+}{}_{(aq)} + 2Cl^{-}{}_{(aq)} \rightarrow Sr(OH)_{2(aq)} + 2NaCl_{(aq)} \qquad (iii)$$

$$2Sr(OH)_{2(aq)} \rightarrow 2SrO_{(s)} \downarrow + 2H_2O_{(aq)} \qquad (iv)$$

$$SrO + CNT(Disperssed) \rightarrow SrO.CNT_{(s)} \downarrow \qquad (v)$$

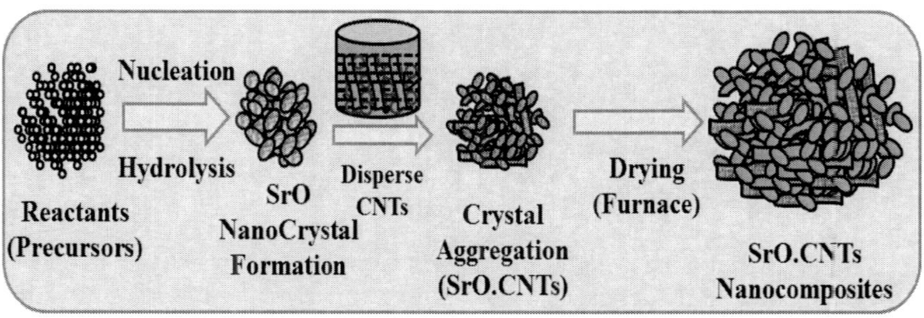

Figure 1. Growth mechanism of SrO.CNT NCs by wet-chemical process.

Electrode Fabrication with SrO.CNT NCs

An electrode is an electrical conductor used to make contact with a nonmetallic part of a circuit (e.g., a semiconductor, an electrolyte, a vacuum or air). The word was coined by William Whewellat the request of the scientist Michael Faraday from the Greek word s*elektron*, meaning amber (from which the word electricity is derived), and *hodos*, a way. An electrode in an electrochemical cell is referred to as either an *anode* or a *cathode*. The anode is now defined as the electrode at which electrons leave the cell and oxidation occurs, and the cathode as the electrode at which electrons enter the cell and reduction occurs. Each electrode may become either the anode or the cathode depending on the direction of current through the cell. A bipolar electrode is an electrode that functions as the anode of one cell and the cathode of another cell. In this research approach, it was used two electrode system by I-V method.

(a) Glassy carbon electrode: Glass-like carbon, often called glassy carbon or vitreous carbon, is a non-graphitizing, or non-graphitie carbon which combines glassy and ceramic properties with those of graphite. The most important properties are high temperature resistance, hardness (7Mohs), low density, low electrical resistance, low friction, low thermal resistance, extreme resistance to chemical attack and impermeability to gases and liquids. Glassy carbon is widely used as an electrode material in electrochemistry, as well as for high temperature crucibles and as a component of some prosthetic device s, and can be fabricated as different shapes, sizes and sections. The names *glassy carbon* and *vitreous carbon* have been introduced as trademarks; therefore, IUPAC does not recommend their use as technical terms. Vitreous carbon can also be produced as foam. It is then called reticulated vitreous carbon (RVC). This foam was first developed in the mid to late 1960s as a thermally insulating, micro-porous glassy carbon electrode material. RVC foam is a strong, inert, electrically and thermally conductive, and corrosion resistant porous form of carbon with a low resistance to gas and fluid flow. Due to these characteristics, the most widespread scientific use of RVC is in electrochemistry. Additionally, RVC foams are characterized by an exceptionally high void volume, high surface area, and very high thermal resistance in non-oxidizing environments, which allows for heat sterilization and facilitates manipulation in biological applications.

(b) Gold electrode: The working electrode is the electrode in an electrochemical system on which the reaction of interest is occurring. The working electrode is often used in conjunction with an auxiliary electrode, and a reference electrode in a three electrode system. Depending on

whether the reaction on the electrode is a reduction or an oxidation, the working electrode is called cathodic or anodic, respectively. Common working electrodes can consist of materials ranging from inert metals such as gold, silver or platinum, to inert carbon such as glassy carbon or pyrolytic carbon, and mercury drop and film electrodes. Chemically modified electrodes are employed for the analysis of both organic and inorganic samples.

(c) Counter electrode: The auxiliary electrode, often also called the counter electrode, is an electrode used in a three electrode electrochemical cell for voltammetric analysis or other reactions in which an electric current is expected to flow. The auxiliary electrode is distinct from the reference electrode, which establishes the electrical potential against which other potentials may be measured, and the working electrode, at which the cell reaction takes place. In a two-electrode system, both a known current or potential is applied between the working and auxiliary electrodes and the other variable may be measured. The auxiliary electrode functions as a cathode whenever the working electrode is operating as an anode and vice versa. The auxiliary electrode often has a surface area much larger than that of the working electrode to ensure that the half-reaction occurring at the auxiliary electrode can occur fast enough so as not to limit the process at the working electrode. When a three electrode cell is used to perform electro analytical chemistry, the auxiliary electrode, along with the working electrode, provides circuit over which current is either applied or measured. Here, the potential of the auxiliary electrode is usually not measured and is adjusted so as to balance the reaction occurring at the working electrode. This configuration allows the potential of the working electrode to be measured against a known reference electrode without compromising the stability of that reference electrode by passing current over it. The auxiliary electrode may be isolated from the working electrode using a glass frit. Such isolation prevents any byproducts generated at the auxiliary electrode from contaminating the main test solution: for example, if a reduction is being performed at the working electrode in aqueous solution, oxygen may be evolved from the auxiliary electrode. Such isolation is crucial during the bulk electrolysis of a species which exhibits

reversible redox behavior. Auxiliary electrodes are often fabricated from electrochemically inert materials such as gold, platinum, or carbon. In this approach, Na_2HPO_4 (0.2 M), NaH_2PO_4 (0.2 M) and distilled water (100.0 mL) were used in the preparation of phosphate buffer (PB, 0.1 M, pH = 7). GCE (surface area = 0.0316 cm^2) were cleaned thoroughly with acetone, and placed at open air to dry (1.0 h). NMs (SrO NPs, and SrO.CNT NCs) were then dispersed ethanol, subsequently deposited on the dried surface of GCE, and again placed at open air to dry (2 h). After drying, conducting binding agent, nafion was added in drop wise with the deposited NMs, and kept in open air until completed drying with uniform film formation. The fabricated NMs and platinum wire (Pt) were used as working and counter electrode respectively.

RESULTS AND DISCUSSION

Study of Optical and Structural Property

Ultraviolet–visible spectroscopy or ultraviolet-visible spectrophotometry (UV-Vis) refers to absorption spectroscopy or reflectance spectroscopy in the ultraviolet-visible spectral region. This means it uses light in the visible and adjacent (near-UV and near infrared, NIR ranges). The absorption or reflectance in the visible range directly affects the perceived color of the chemicals involved. In this region of the electromagnetic spectrum, molecules undergo electronic transitions. This technique is complementary to fluorescence spectroscopy, in that fluorescence deals with transitions from the excited state to the ground state, while absorption measures transitions from the ground state to the excited state [110, 111]. In principle molecules containing π-electrons or non-bonding electrons (n-electrons) can absorb the energy in the form of ultraviolet or visible light to excite these electrons to higher anti-bonding molecular orbital [112, 113]. The more easily excited the electrons (i.e., lower energy gap between the HOMO and the LUMO), the longer the wavelength of light it can absorb. For applications, UV-Vis spectroscopy is routinely used in analytical chemistry for the quantitative determination of different analytes, such as transition metal ions, highly conjugated organic compounds, and biological macromolecules. The optical property is one of the prime criteria for the assessment of photo-catalytic activity of the grey grown SrO.CNT NCs. According to the UV-Visible principle, the outer electrons of the atom absorb radiant energy and make transitions to high energy levels. Due to the optical absorption, the spectrum including band-gap energy of the metal oxide was obtained. The UV-Vis spectrum of the SrO NPs and SrO.CNT NCs solution were recorded in the visible range (200-800 nm) and a broad absorption band at around 296.8 and 320 nm were found (Figure 1a-c). Based on the maximum level band absorption, the band-gap energy of the SrO NPs and SrO.CNT NCs were calculated using equation (v), and according to the direct band-gap rule [Tauc's equation, (vi)], curve of $(\alpha h\nu)^2 vs h\nu$ were plotted and then extrapolated to the x-axis. From the extrapolated curve, the band-gap energy of the SrO NPs and SrO.CNT NCs were found as~ 3.0 and 2.9 eV respectively (Figure 1b-d). Where, BE = Band gape energy, λ_{max} = Maximum absorption wavelength, α = Absorption coefficient, A = Constant related to the effective mass of the electrons, r = 0.5 (Direct transition), h = Plank's constant, ν = Frequency. There was no additional peaks were found in the spectrum associated with impurities and structural defects, which recognized that the crystalline nature of the prepared SrO NPs and SrO.CNT NCs [43, 92].

$$BE = \frac{1240}{\lambda_{max}} \qquad (vi)$$

$$(\alpha h\nu)^{1/r} = A(h\nu - BE) \qquad (vii)$$

Fourier transform infrared spectroscopy (FTIR) [114] is a technique which is used to obtain an infrared spectrum of absorption or emission of a solid, liquid or gas. An FTIR spectrometer simultaneously collects high spectral resolution data over a wide spectral range. This confers a significant advantage over a dispersive spectrometer which measures intensity over a narrow range of wavelengths at a time. The term *Fourier transform infrared spectroscopy* originates from the fact that a Fourier transform (a mathematical process) is required to convert the raw data into the actual spectrum. The goal of any absorption spectroscopy (FTIR, ultraviolet-visible ("UV-Vis") spectroscopy, etc.) is to measure how well a sample absorbs light at each wavelength. The processing required turns out to be a common algorithm called the Fourier transform (hence the name, "Fourier transform spectroscopy"). The raw data is sometimes called an "interferogram."The first low-cost spectrophotometer capable of recording an infrared spectrum was the Perkin-Elmer Infra cord produced in 1957. This instrument covered the wavelength range from 2.5 μm to 15 μm (wavenumber range 4000 cm^{-1} to 660 cm^{-1}). The lower wavelength limit was chosen to encompass the highest known vibration frequency due to a fundamental molecular vibration. The upper limit was imposed by the fact that the dispersing element was a prism made from a single crystal of rock-salt (sodium chloride) which becomes opaque at wavelengths longer than about 15 μm; this spectral region became known as the rock-salt region. Later instruments used potassium bromide prisms to extend the range to 25 μm (400 cm^{-1}) and caesium iodide50 μm (200 cm^{-1}). The region beyond 50 μm (200 cm^{-1}) became known as the far-infrared region, at very long wavelengths it merges into the microwave region. Measurements in the far infrared needed the development of accurately ruled diffraction gratings to replace the prisms as dispersing elements since salt crystals are opaque in this region. More sensitive detectors than the bolometer were required because of the low energy of the radiation. One such was the Golay detector. An additional issue is the need to exclude atmospheric water vapor because water vapor has an intense pure rotational spectrum in this region. Far-infrared spectrophotometers were cumbersome, slow and expensive. The grey grown SrO.CNT NCs were also characterized in perspective of atomic and molecular vibrations to determine the functional properties and in this regards, FT-IR spectrum was recorded in the region of 4000-400 cm^{-1} under normal condition. The FT-IR spectrum of the SrO NPs and SrO.CNT NCs (Figure 2a) shows peaks at 3282 (s, br), 2105 (w), 1552 (w), 1354 (m), 1103 (s), 852 (m) and 605 (m) cm^{-1} which denoted the presence of O-H (stretching), −C≡C− (stretching), −C=C− (stretching), C−H (rocking), Sr−O−Sr (stretching), C−H and Sr=O (stretching) respectively in the NCs [115]. The observed peak at 605 cm^{-1} recognized the formation of metal-oxide (Sr=O) bond that indicates the configuration of the SrO NPs and SrO.CNT NCs.

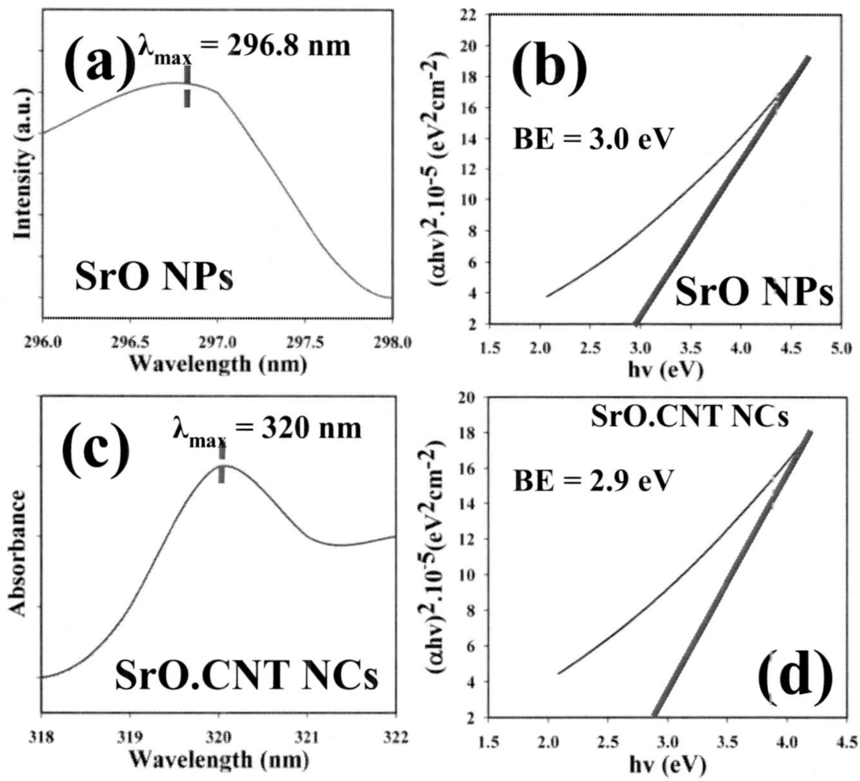

Figure 2. (a-c) UV-Vis spectra, (b-d) Band-gap energy plot of SrO NPs and SrO.CNT NCs respectively.

X-ray crystallography is a tool used for identifying the atomic and molecular structure of a crystal, in which the crystalline atoms cause a beam of incident X-rays to diffract into many specific directions. By measuring the angles and intensities of these diffracted beams, a crystallographer can produce a three-dimensional picture of the density of electrons within the crystal. From this electron density, the mean positions of the atoms in the crystal can be determined, as well as their chemical bonds, their disorder and various other information. Since many materials can form crystals such as salts, metals, minerals, semiconductors, as well as various inorganic, organic and biological molecules X-ray crystallography has been fundamental in the development of many scientific fields. In its first decades of use, this method determined the size of atoms, the lengths and types of chemical bonds, and the atomic-scale differences among various materials, especially minerals and alloys. The method also revealed the structure and function of many biological molecules; including vitamins, drugs, and protein sand nucleic acid s such as DNA. X-ray crystallography is still the chief method for characterizing the atomic structure of new materials and in discerning materials that appear similar by other experiments. X-ray crystals structures can also account for unusual electronic realistic properties of a material, shed light on chemical interaction s and processes, or serve as the basis for designing pharmaceuticals against diseases. XRD analysis was performed to study the crystal pattern of the prepared SrO NPs and SrO.CNT NCs. The found potential peaks with indication for 2θ values at 21.5, 25 (100), 30, 32 (110), 35, 37.5, 42 (111), 44, 46 (200), 47, 48, 50, 56 (210), 57, 58 (211), 60, 64, 66 (220), 72 (221), 74 and 78 (310) degrees (Figure 2b).

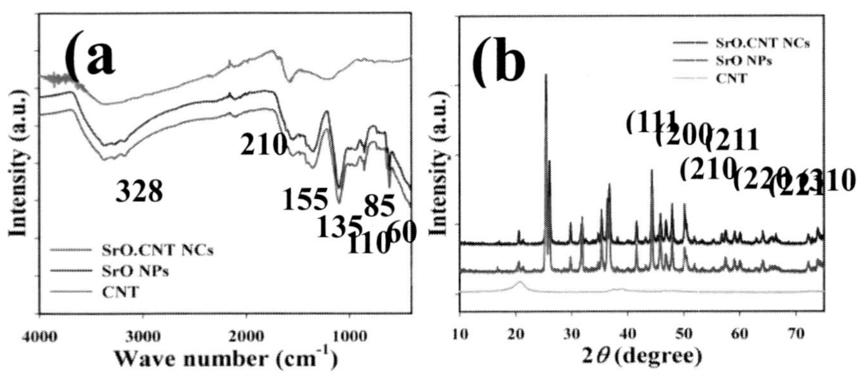

Figure 3. (a) FT-IR spectra and (b) XRD patterns of SrO.CNT NCs.

All the observed peaks in the spectrum were assigned by using the JCPDS file (No: 001-1113, 6-520, 75-0263 and 01-073-0661). The strong peaks have been denoted the crystalline nature and purity of the NCs. The crystalline pattern is the indication of metal-oxygen framework and from these parameters we can say that a significant amount of crystalline SrO present in the doped NCs [107, 116-118].

Evaluation of Morphological and Elemental Analyses

As scanning electron microscope (SEM) is a type of electron microscope that produces images of a sample by scanning it with a focused beam of electrons. The electrons interact with atoms in the sample, producing various signals that contain information about the sample's surface topography and composition. The electron beam is generally scanned in arrester scan pattern, and the beam's position is combined with the detected signal to produce an image. SEM can achieve resolution better than 1.0 nm. Specimens can be observed in high vacuum, in low vacuum, in wet conditions (in environmental SEM), and at a wide range of cryogenic or elevated temperatures. The most common SEM mode is detection of secondary electrons emitted by atoms excited by the electron beam. The number of secondary electrons that can be detected depends, among other things, on the angle at which beam meets surface of specimen, i.e., on specimen topography. By scanning the sample and collecting the secondary electrons that are emitted using a special detector, an image displaying the topography of the surface is created. Back scattered electron imaging, quantitative X-ray analysis, and X-ray mapping of specimens often requires that the surfaces be ground and polished to an ultra-smooth surface. Specimens that undergo Wavelength dispersive spectrometers (WDS) or Energy dispersive spectrometers (EDS) analysis are often carbon coated. In general, metals are not coated prior to imaging in the SEM because they are conductive and provide their own pathway to ground. In general microscopy information concerns surface crystallography (i.e., how the atoms are arranged at the surface), surface morphology, and surface composition. In FEM, the phenomenon of field electron emission was used to obtain an image on the detector on the basis of the difference in work function of the various crystallographic planes on the surface. FESEM is one of the perfect techniques to examine the morphology of NCs including nanostructure materials. The morphology and elemental properties of the grey grown prepared SrO.CNT NCs were investigated using FESEM and EDS respectively. Using FESEM, the typical shapes of CNT, SrO NPs and SrO.CNT NCs were recorded at low to high magnified region (Figure 3a-d). The magnified images indicated that SrO was aggregated with a bright contrast including well dispersed on the surface of CNTs (Figure 3c-d). The conductance of NCs may be increased with the incorporation of SrO onto CNT, which correlated the calculation of band-gap energy (BE) of two compounds.

Energy-dispersive X-ray spectroscopy (EDS, EDX, or XEDS), sometimes called energy dispersive X-ray analysis (EDXA) or energy dispersive X-ray microanalysis (EDXMA), is an analytical technique used for the elemental analysis or chemical characterization of a sample. It relies on an interaction of some source of X-ray excitation and a sample. Its characterization capabilities are due in large part to the fundamental principle that each element has a unique atomic structure allowing unique set of peaks on its X-ray emission spectrum [119].To stimulate the emission of characteristic X-rays from a specimen, a high-energy beam of charged particles such as electrons or protons, or a beam of X-rays, is focused into the sample being studied. At rest, an atom within the sample contains ground state (or unexcited) electrons in discrete energy levels or electron shells bound to the nucleus. The incident beam may excite an electron in an inner shell, ejecting it from the shell while creating an electron hole where the electron was. An electron from an outer, higher-energy shell then fills the hole, and the difference in energy between the higher-energy shell and the lower energy shell may be released in the form of an X-ray. The number and energy of the X-rays emitted from a specimen can be measured by an energy-dispersive spectrometer. As the energies of the X-rays are characteristic of the difference in energy between the two shells and of the atomic structure of the emitting element,

EDS allows the elemental composition of the specimen to be measured. On the basis of EDS analysis, carbon (C), oxygen (O) and strontium (Sr) were present in the prepared grey grown SrO.CNT NCs and consists of C (21.60), O (21.44) and Sr (56.96) wt% respectively (Figure 4a-d). According to the FESEM coupled with EDS, no other peaks related with impurities were observed which indicated the NCs are composed of C including O and Sr. A comparison among CNT, SrO NPs and SrO.CNT NCs in perspective of weight (%) is given in Table 2.

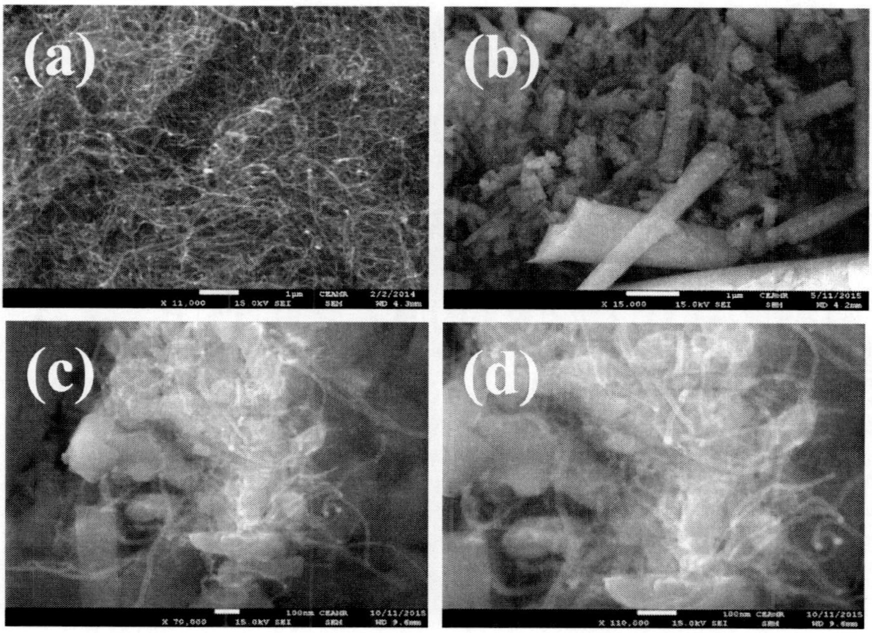

Figure 4. FESEM magnified images (a) CNT, (b) SrO NPs (c-d) SrO.CNT NCs.

Table 1. Weight (%) of the elements presented in the NMs

Materials	Weight (%)		
	C	O	Sr
CNT	94.55	-	-
SrO NPs	-	27.35	72.65
SrO.CNT NCs	21.60	21.44	56.96

Table 2. Calculated binding energy from XPS observation

Materials	Binding energies (eV)		
CNT	Sr^{2+}	C1s	O1s
SrO NPs	-	285.0	-
SrO.CNT NCs	133.2, 139.0	-	537.0

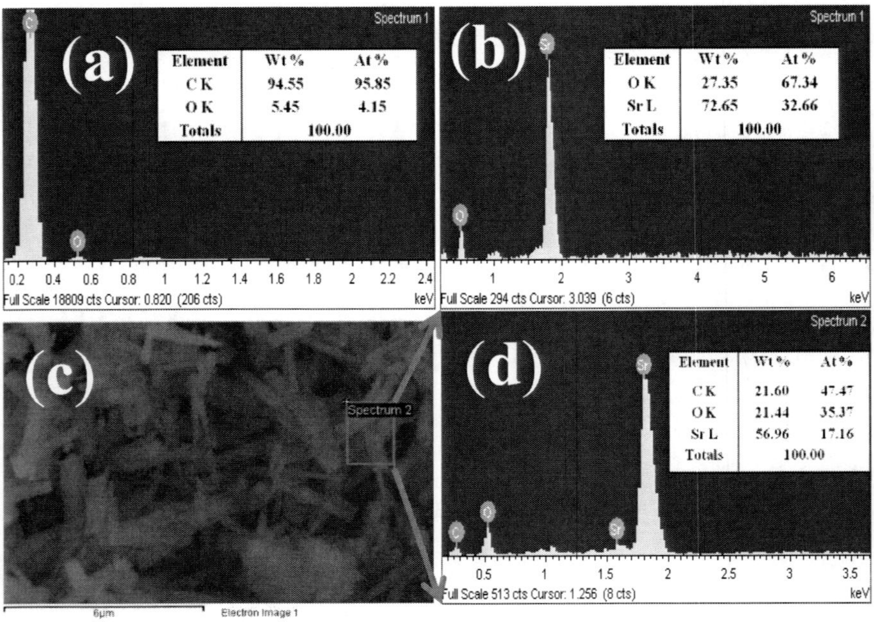

Figure 5. Elemental analysis: (a) CNT, (b) SrO NPs and (c-d) SrO.CNT NCs.

Study of XPS for Binding Energy

X-ray Photoelectron Spectroscopy (XPS) also known as Electron Spectroscopy for Chemical Analysis (ESCA) is the most widely used surface analysis technique because it can be applied to a broad range of materials and provides valuable quantitative and chemical state information from the surface of the material being studied. The average depth of analysis for an XPS measurement is approximately 5 nm. PHI XPS instruments provide the ability to obtain spectra with a lateral spatial resolution as small as 7.5 μm spatial distribution information can be obtained by scanning the micro focused x-ray beam across the sample surface. Depth distribution information can be obtained by combining XPS measurements with ion milling (sputtering) to characterize thin film structures. The information XPS provides about surface layer s or thin film structures is important for many industrial and research applications where surface or thin film composition plays a critical role in performance including: NMs, photo-voltaic, catalysis, corrosion, adhesion, electronic devices and packaging, magnetic media, display technology, surface treatment s, and thin film coatings used for numerous applications. XPS is typically accomplished by exciting a samples surface with mono-energetic Alpha x-rays causing photoelectrons to be emitted from the sample surface. An electron energy analyzer is used to measure the energy of the emitted photoelectrons. From the binding energy and intensity of a photoelectron peak, the elemental identity, chemical state, and quantity of a detected element can be determined. Physical Electronics XPS instruments function in a manner analogous to SEM-EDS instruments that use a finely focused electron beam to create SEM images for sample viewing and point spectra or images for compositional analysis. XPS spectra are obtained by irradiating a material with a beam of X-rays while simultaneously measuring the kinetic energy and number of electrons that escape from the top 0 to 10 nm of the material being analyzed. XPS requires high vacuum ($P \sim 10^{-8}$ millibar) or ultra-high vacuum (UHV; $P< 10^{-9}$millibar) conditions, although a current area of development is ambient-pressure XPS, in which samples are analyzed at pressures of a few tens of millibar. XPS is a surface chemical analysis technique that can be used to analyze the surface chemistry of a material in its as received state, or after some treatment, for example, fracturing, cutting or scraping in air or UHV to expose the bulk chemistry, ion beam etching to clean off some or all of the surface contamination (with mild ion etching) or to intentionally expose deeper layers of the sample (with more extensive ion etching) in depth-profiling XPS, exposure to heat to study the changes due to heating, exposure to reactive gases or solutions. The presence of the chemical nature of the elements present in NCs could be indicated by a quantitative spectroscopic technique, XPS. XPS spectra may be recorded by irradiating of X-ray beam with a NCs material and the kinetic energy including electrons number of the sample can be measured simultaneously. On the basis of XPS spectra, carbon, oxygen and strontium were present in the synthesized SrO.CNT NCs (Figure 5a). The spin orbit of strontium $Sr3d_{5/2}$ and $Sr3d_{3/2}$ denote the peaks at 133.0 and 137.0 eV respectively which indicated that strontium (Sr^{2+}) were present in the NCs (Figure 5b). The C1s and O1s spectra recognize the main peak at 283.0 and 535.0 eV which denoted that carbon (C) and oxygen (O^{2-}) were also present in the prepared SrO.CNT NCs (Figure 5c-d) [118, 121]. A comparison regarding binding energies among CNT, SrO NPs and SrO.CNT NCs is presented in Table 3.

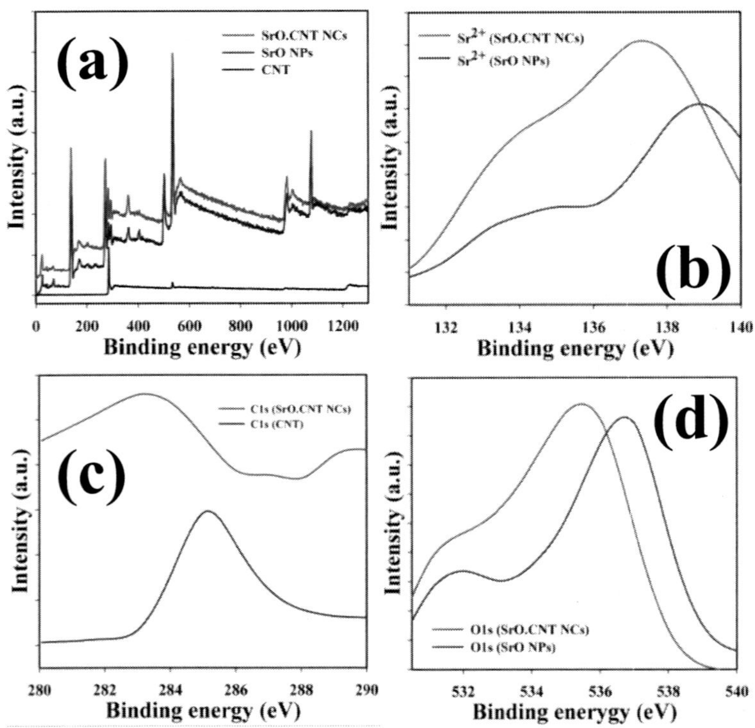

Figure 6. Study of X-ray photoelectron spectroscopy (a) Full spectra (b) Sr $3d_{5/2}$ and Sr $3d_{3/2}$ level, (c) C1s level and (d) O1s level of SrO.CNT NCs, SrO NPs and CNT acquired with MgKα1 radiations. X-ray beam-spot size is 400.0 μm; Pass-energy is 200.0 eV; Pressures less than 10^{-8} Torr.

Table 3. A comparison study of the analytical performances of different electrochemical methods for hydrazine determination

Electrode	Method	Medium	pH	LOD (μM)	LDR (μM)	Sensitivity ($\mu AmM^{-1}cm^{-2}$)	Ref.
HPIMBD/MWCNT/GCE	AM	PB	7.0	1.1	4.0-32.9	1.6×10^{-5}	[64]
MnO_2/GO/GCE	AM	PB	7.0	0.16	-	1.007	[65]
Ag/ZIF-8/CPE	AM	NaOH	-	1.57	6-5000	5.446	[69]
TiO_2-Pt NFs/GCE	AM	PB	7.0	0.142	1.0 mM	4.42	[129]
CM-MWCNT modified GC	AM	PB	8.0	1.4	2-44	2.29	[130]
NiHCF modified graphine	AM	$NaNO_3$	7	1.0	2.4-8200	1.13	[131]
C@ZnO nanorod array	AM	PB	8.0	0.1	0.1-3.8	-	[132]
CGA modified GC	CA	PB	7.5	-	50-11000	-	[133]
PtNPs/TiO_2NSs/GCE	CA + CV	PB	7.0	2.0	20-900	1.874	[62]
AuNPS/PDTYP/MWCNTS/GCE	CV	PB	8.0	0.6	2.0-130	-	[134]
O-AP modified GC	CV	PB	9.0	0.5	2.0-20.0	0.016	[135]
FePc/Au	CV + OSWV	PB	7.0	5.0, 11.0	13-92	1.62×10^{-5}	[136]
Au/PPy/GC	DPV	PB	7.0	0.20	5-200	1.26	[137]
Pd NPS decorated BDD	LSV	PB	7.0	2.6	27.2-85	0.60	[138]
SrO.CNT NCs/GCE	I-V	PB	7.0	0.036 nM	0.2 nM– 0.2 M	26.37	[88]

*BDD: Boron-doped diamond, AM: Amperometry, CA: Chronoamperometry, CGA: Chlorofenic acid, CM: Curcumin, CNTs: Carbon nano tubes, CV: Cyclic voltametry, FePC: Iron Pathalocyamine, DPV: Differential pulse voltametry, HPIMBD: 4-((2-hydroxyphenylimino)methyl)benzene -1,2-diol, MWCNT: Multiwalled carbon nano tubes, NCs: Nanocomposites, NiHCF: Nickel hexacyanoferrate, NFs: Nano fibers, NPs: Nanoparticles, NSs: Nano sheets, OSWV: Ostero young square wave voltametry, PPy: Polypyrrole, PB: Phosphate buffer solution, and LSV: Linear sweep voltametry.

SENSOR APPLICATIONS

Detection of Hydrazine by SrO.CNT NCs

Electrochemical methods comprise a collection of extremely useful measurement tools for neuroscience. A central feature of these methods is an electrode that provides a surface or interface where some form of a charge-transfer process occurs. This charge transfer process gives rise to potentials and/or currents that can be measured and related either by theory or by calibration to the concentration of substances in the solution that bathes the electrode. Broadly speaking, these methods can be divided into two groups: measurements that do not involve current, known generally as potentio-metric methods, and measurements that involve current flow at an electrode under potential control, known as amperometry, voltammetry, or polarography, depending upon the details of the experimental design. It is completely natural that neuroscientists should find electrochemical methods of value because the nervous system contains numerous electrochemically detectable targets including simple inorganic ions, catecholamine and indolamine neurotransmitters and their metabolites, glutamate, nitric oxide (NO), glucose, lactate, ascorbate, urate, hydrogen peroxide (H_2O_2), oxygen (O_2), and pH. Moreover, the high ionic strength of biological environments creates a perfect milieu for electrochemistry, which requires conductive media.

The invention of the carbon fiber microelectrode by Gonon and colleagues was a milestone in the development of significant electrochemical methods in neuroscience. These electrodes, which are still popular, could be electrochemically pretreated to allow resolution of the oxidation currents recorded in the brain into distinct peaks at different positions along the potential axis. These studies demonstrated the first of numerous approaches to the development of a variety of selective electrochemical recording techniques that are currently used in the laboratories of the contributors to this volume. Another major milestone in the development of selective recording techniques was the introduction of carbon electrodes coated with the perm-selective polymer material called

Nafion, which is a sulfonated derivative of Teflon. The sulfonate groups ionize in wet Nafion to produce fixed anionic sites in the membrane. These fixed anionic sites limit the ability of anionic substances in the brain from reaching the electrode surface, which has proven very useful as a way to eliminate signal contributions from anionic substances such as ascorbate, urate, and the acidic metabolites of the monoamine neurotransmitters. Hence, Nafion increases the selectivity of the electrochemical techniques for cationic substances such as the monoamine neurotransmitters themselves. In addition, the Teflon-like nature of Nafion imparts the electrode surface with the nonstick characteristics that make Teflon itself of interest in cookware. Nafion seems to improve the biocompatibility of the electrode surface and diminish fouling of the electrodes during in vivo experiments.

The potential application of SrO-CNT NCs assembled onto GCE as chemical sensor (especially hydrazine analyte in buffer system) has been executed for measuring and detecting target chemical. Enhancement of the SrO.CNT NCs/GCE as chemical sensors is in the initial stage and no other reports are available. The SrO.CNT NCs/GCE sensor has advantages such as stability in air, non-toxicity, chemical inertness, electro-chemical activity, simplicity to assemble, ease in fabrication, and chemo-safe characteristics. As in the case of hydrazine sensors, the current response in I-V method of SrO.CNT NCs/GCE considerably changes when aqueous hydrazine is adsorbed. The SrO.CNT NCs/GCE was applied for fabrication of chemi-sensor, where hydrazine was measured as target analyte. The fabricated surface of SrO.CNT NCs/GCE sensor was prepared with conducting binder (5% ethanolic solution) on the GCE surface. I-V signals of hydrazine chemical sensor are anticipated having SrO.CNT NCs/GCE on thin-film as a function of current versus potential. The resultant electrical responses of hydrazine are investigated by simple and reliable I-V technique using SrO.CNT NCs/GCE. The holding time of electrometer was set for 1.0 sec. A significant amplification in the current response with applied potential is noticeably confirmed. The simple and possible reaction mechanism is generalized in Figure 6 in presence of SrO.CNT NCs/GCE surfaces by I-V method. In presence of SrO.CNT NCs/GCE, the electrons are released in presence of hydrazine by adsorbing reduced oxygen, which improved and enhanced the current responses against potential during the I-V measurement at room conditions. According to the previous reports the electrochemical reaction for the electro-catalytic oxidation of hydrazine (Reaction ix) is proposed by Hu et al., [120].

$$N_2H_4(SrO.CNTNCs) + \tfrac{5}{2}OH^- \rightarrow \tfrac{1}{2}N^{-3} + \tfrac{1}{2}NH_3 + \tfrac{5}{2}H_2O + 2e^- \quad (viii)$$

The NCs materials have been reported earlier as chemical sensors [121-124]. The significant application of SrO.CNT NCs assembled onto an electrode as chemical sensor has been employed for the detection of materials that are not environmentally toxic. The SrO.CNT NCs sensors posses lots of typical advantages including biologically safe, chemical stability, consistency in air, high electrochemical activity, nontoxicity, large surface area and easy to assemble.

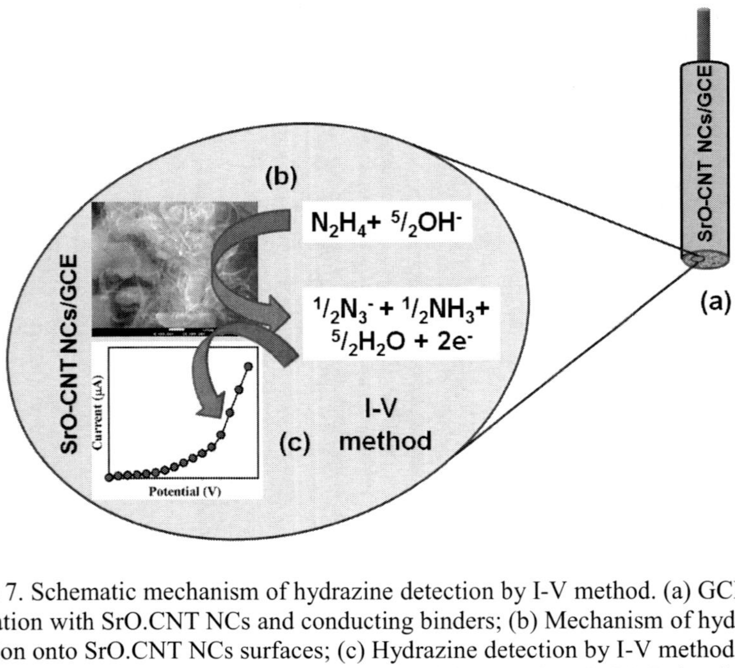

Figure 7. Schematic mechanism of hydrazine detection by I-V method. (a) GCE fabrication with SrO.CNT NCs and conducting binders; (b) Mechanism of hydrazine detection onto SrO.CNT NCs surfaces; (c) Hydrazine detection by I-V method using SrO.CNT NCs modified GCE. PB: 0.1 M; Potential range: 0~ +1.5V; Delay time: 1s; Surface area of GCE: 0.0316 cm^2; Method: I-V.

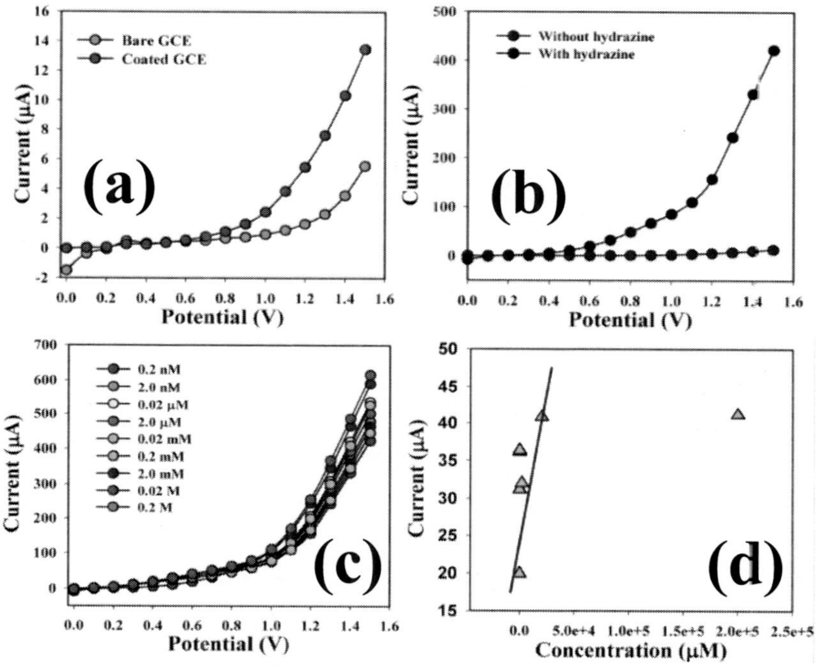

Figure 8. (a) Uncoated and coated electrode with SrO.CNT NCs, (b) Absence and presence of hydrazine with SrO.CNT NCs electrode, (c) Concentration variation of the hydrazine and (d) Calibration curve of the SrO.CNT NCs fabricated on GCE surface.

The current response in the I-V method for the SrO.CNT NCs considerably changed during adsorption of hydrazine as target agent. According to the potential range (0.1 to 1.5 V), the current responses for the uncoated bare GCE and coated with SrO.CNT NCs on the working electrode surface were presented in Figure 6a. From the Figure 7a, it was revealed that the differences of the current responses between bare and coated GCE and it was occurred due to the current signals were more affected by coated electrode in compared with bare GCE. The target molecule, hydrazine (0.2 nM ~ 0.2 M) was added to the SrO.CNT NCs modified electrode and the changes of current responses were recorded for without (blue-dotted) and with (black-dotted) analyte (Figure 7b). Due to the presence of NCs, a significant enhancement of current was achieved with hydrazine which gives a higher surface area with better coverage in absorption and adsorption capability onto the porous NCs surfaces of the target molecule (hydrazine). The responses (current *vs* voltage) of the SrO.CNT NCs modified electrode were recorded for the different concentration (0.2 nM ~ 0.2 M) of the hydrazine which indicates that the changes of current of the fabricated electrode as a function of hydrazine concentration under normal condition (Figure 7c) and it was also reported that the current responses increased regularly from lower to higher concentration of the target agent. For determination of the probable analytical limit, a broad range of the analyte concentrations (0.2 nM ~ 0.2 M) were measured from the lower to higher potential (0.1 to 1.5 V). From the various concentrations of hydrazine, the linear calibration and magnified calibration curve at 0.5 V were plotted (Figure 7d). The linear dynamic range (0.2 nM-0.2 M), regression coefficient (r^2 = 0.7311), detection limit (0.036 ± 0.02 nM) and sensitivity (~26.37 $\mu AmM^{-1}cm^{-2}$) at signal to slope ratio = 3 were calculated from the calibration curve.

The resistance value of the SrO.CNT NCs modified GCE chemical sensors can be decreased with enhancing active surface area is an important criterion of the NCs particles [125, 126]. These reactions are consummated in bulk-system/air-liquid interface /neighboring atmosphere owing to the small carrier concentration, which increase the resistance. The hydrazine sensitivity towards SrO.CNT NCs can be ascribed to the higher-oxygen lacking conducts to enhance the oxygen adsorption. Larger the amount of oxygen adsorbed on the SrO-doped NCs-sensor surface, higher would be the oxidizing potentiality and faster would be the oxidation of hydrazine. The activity of hydrazine would have been extremely big as contrast to other toxic chemical with the surface under indistinguishable conditions [127, 129]. The selectivity was performed with different chemicals such as 2-nitrophenol, 4-aminophenol, 4-methoxyphenol, acetone, benzaldehyde, bisphenol A, ethanol, hydrazine, melamine and ammonium hydroxide (Figure 8a). Hydrazine showed maximum current responses to SrO.CNT NCs fabricated electrode and therefore it was clearly notified that the sensor was most selective towards hydrazine compared with other chemicals. Response time was measured at 2.0 μM and found 7.0 s (Figure 8b). I-V response of the SrO.CNT NCs coated electrode sensor was performed up to 2 weeks for the determination of the reproducibility and storage stabilities. It was marked that the current response was not significantly changes after washing of each experiment of the fabricated SrO.CNT NCs electrode substrate (Figure 8c). The sensitivity remained almost similar as the initial value up to two week and after that the responses of the fabricated electrode become decreased gradually. At different conditions, a series of seven successive measurements of hydrazine solution (0.2 μM) yielded good reproducible responses with the SrO.CNT NCs electrode. A control experiment was also conducted at hydrazine concentration (0.2 μM) with different fabricated electrodes and a significant increase of current response was marked for the SrO.CNT NCs compared with SrO (Figure 8d).

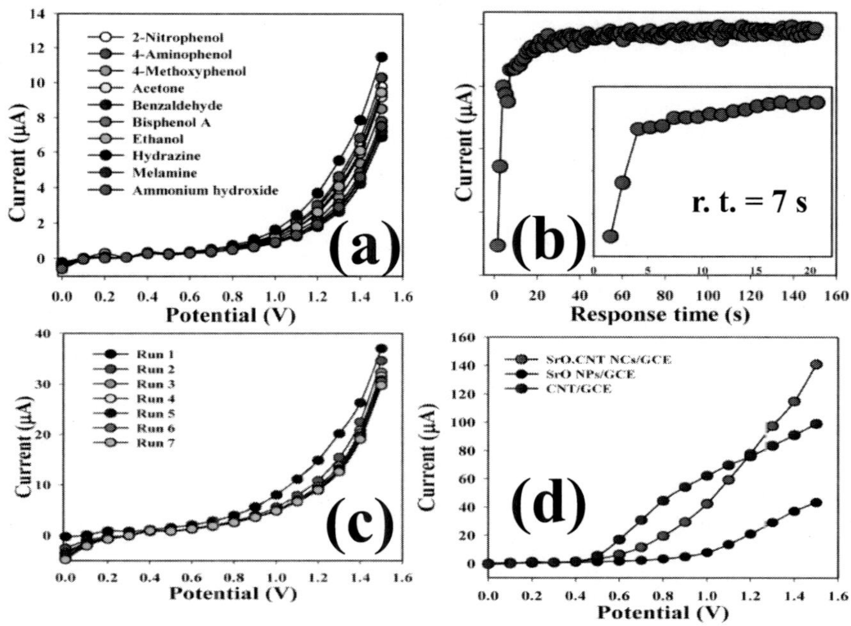

Figure 9. SrO.CNT NCs coated electrode for hydrazine sensing I-V responses: (a) selectivity, (b) response time, (c) reproducibility study and (d) control experiment.

The responses of NCs sensor were determined with respect to storage time for measurement of long term storage stability. The storage stability measurement of the SrO.CNT NCs electrode sensor was conducted under normal conditions and the sensitivity remained almost 90% for several days as the initial responses. From the above experiment it was clearly revealed that the fabricated sensor might be used without any significant deterioration of sensitivity up to several weeks. Response time and limit of dynamic response (0.2 nM ~ 0.2 M) of the electrode sensor were reported from the practical concentration variation graph. The SrO.CNT NCs unusual regions dispersed on the surface could improve the ability of CNT NCs to absorb more O_2 species giving higher resistances in PB system. In previous studies hydrazine was determined using different chemical techniques. Here, we report the determination of hydrazine using current *vs* voltage method. The analytical performances of different electrochemical procedures for the determination of hydrazine are given in Table 4.

Study of Interference Effect (IE)

IE evaluation is one of the important methods in analytical science, owing to the ability to differentiate the interfering constituents having similar chemical background in the chemical sensor [73]. Phenyl hydrazine (PHy), 4-Nitrophenyl hydrazine (4-NPHy), 3-Methoxyphenyl hydrazine (3-MPHy), and *m*-Tolyl hydrazine hydrochloride (m-THyHCl) are generally used as interfering chemicals in the hydrazine chemical sensor. I-V responses at SrO.CNT NCs/GCE toward the addition of Hydrazine (Hy), and interfering chemicals such as PHy, 4-NPHy, 3-MPHy and m-THyHCl (2.0 μM) in PB (pH = 7.0) were recorded. The effects of interfering chemicals towards hydrazine sensor were calculated at calibrated potential (+ 0.5 V), and compared with Hy effect (Figure 9a-b and Table 4), where IE effect of Hy effect was considered to be 100 %. From the IE experiment, it was clear that SrO.CNT NCs/GCE did not reveal any significant responses towards interfere chemicals. So, the projected sensor is suitable for detection of hydrazine with good sensitivity.

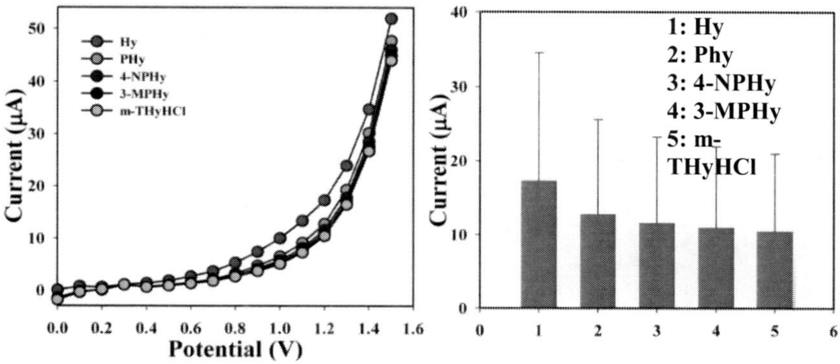

(a)(100) **(b)(110)**

Figure 10. Interference effect examination (a) Chemicals and (b) Bar diagram presentation of IE effect at + 1.2 V.

Table 4. Interference effect examination

Chemicals	Current (μA)				IE	SD	RSD
	R1	R2	R3	Average	(%)	(n = 3)	(n = 3)
Hydrazine	0.95	2.68	1.94	1.86	100	1.86	46.75
PHy	0.74	1.19	1.17	1.03	55	0.25	24.60
4-NPHy	0.67	1.13	1.00	0.93	50	0.24	25.41
3-MPHy	0.64	1.11	0.97	0.91	49	0.24	26.52
m-THyHCl	0.58	1.16	0.93	0.89	48	0.29	32.32

Real Sample Analyses

The real samples (industrial effluent, PC bottle safa, PVC food packaging bag, and red sea and tape water) were analyzed in order to validate the proposed I-V method using SrO.CNT NCs/GCE. A standard addition method has been applied to determine the concentration of hydrazine in real samples. A fixed amount (~25.0 μL) of each sample was analyzed in PB (10.0 mL, 0.1 M) using the fabricated SrO.CNT NCs/GCE. The results have been included regarding the quantity of Hy in industrial effluent, PC bottle safa, PVC food packaging bag, sea and tap water samples, which apparently entrenched that the proposed I-V technique is satisfactory, reliable, and suitable for analyzing real samples with assembled of SrO.CNT NCs/GCE (Table 5).

Table 5. Measured Hy concentration in different real samples

Real samples	Calibrated concentration range	Measured Current(μA)	Respective Concentration(nM)
Industrial effluent		3.7	~0.04 ± 0.02
PC bottle safa		3.4	~0.03 ± 0.02
PVC food packaging bag	0.2 nM ~ 200 mM	4.7	~0.05 ± 0.02
Sea water		5.9	~0.06 ± 0.02
Tap water		9.5	~0.10 ± 0.02

SCREENING OF NCS FOR ANTIMICROBIAL PROPERTIES

SrO.CNT NCswere screened for antimicrobial property using disc diffusion method [16]. At known concentration (μg/mL), the solution of the test samples was made by dissolving in definite volume of distilled water (mL). The known amounts of the test samples were impregnated in the dried and sterilized filter paper discs (6 mm) using micropipette. The test microorganisms were seeded uniformly on nutrient agar medium in the discs having test substances. Standard antibiotic and blank discs were used as a positive and negative control respectively and for maximum diffusion, the discs were kept for 24 h at 4°C temperatures. Diffusion was held and for maximum growth of the microorganisms, the plates were then incubated for 24 h at 37°C temperatures according to Barry, 1980 [13]. The growth of microorganisms inhibited by NCs having antimicrobial activity, a clear and distinct zone was observed in the discs. This screening was performed three times and the antimicrobial activity was determined by measuring the diameter of zone of inhibition (mm) of the test materials. The SrO.CNT NCs (100 μg/disc) were screened for antimicrobial properties against two gram positive, gram negative bacteria along with fungi by using standard tetracycline (30 μg/disc) and ketoconazole (30 μg/disc) respectively but no activities was observed (Table 6).

Table 6. Antimicrobial properties of the SrO.CNT NCs

Microbial strains	Zone of inhibition (mm)	
	SrO.CNT NCs	Standard
Gram positive		
Bacillus megaterium	0	50
Staphylococcus aureus (ATCC-25923)	0	35
Gram negative		
Escherichia coli (ATCC-8739)	0	34
Pseudomonas aeruginosa (ATCC-27833)	0	33
Fungi		
Aspergillus niger	0	38
Aspergillus flavus	0	39

CONCLUSION

Finally, SrO.CNT NCs were prepared using active reducing agents through a wet-chemical method, which was cost effective, easy, efficient and simple. The electrochemical characteristics of the NCs were recorded using FTIR, FESEM, EDS, XPS, XRD and UV-Vis instruments. A simple fabrication method used to fabricate SrO.CNT NCs electrode by chemical coating. The sensitive and selective of hydrazine sensor prepared successfully based on electrode embedded with SrO.CNT NCs having conducting binders. The electrochemical parameters of the fabricated hydrazine sensors were good in view of LOD, LDR, sensitivity and short response time. The SrO.CNT NCs electrode recognized higher sensitivity (26.37 $\mu AmM^{-1}cm^{-2}$) and LOD (0.036 nM). A well-established technique can be introduced form this novel approach for the development of efficient chemical sensor for biological and environmental toxic agents in a broad field.

REFERENCES

[1] Adekunle, A., Farah, J., Pillay, K., Ozoemena, B. and Agboola, B. (2012). Electrocatalytic properties of prussian blue nanoparticles supported on poly(m-aminobenzenesulphonic acid)-functionalised single-walled carbon nanotubes towards the detection of dopamine. *Colloids Surfaces B: Biointerfaces,* 95: 186-194.

[2] Adekunle, K., Ozoemena, B. and Agboola, B. (2013). MWCNTs/metal (Ni, Co, Fe) oxide nanocomposite as potential material for supercapacitors application in acidic and neutral media. *J. Solid. State Electrochem.* 17: 1311-1320.

[3] J. Shuskus. (1964). Paramagnetic resonance of some S-state ions in single crystals of strontium oxide. *J. Chem. Phys*. 41:1885.

[4] Mahapatra, K., Maji, R., Maiti, K., Manna, S. K., Mondal, S., Ali, S. S., Manna, S., Sahoo, P., Mandal, S., Uddin, M. R. and Mandal, D. (2015). A BODIPY/pyrene-based chemodosimetric fluorescent chemosensor for selective sensing of hydrazine in the gas and aqueous solution state and its imaging in living cells. *RSC Adv.* 5: 58228-58236.

[5] Mehta. (2011). Ultraviolet-Visible (UV-Vis) Spectroscopy|Analytical Chemistry|Pharmaxchange. Info, Pharmaxchange.

[6] Fakhari, R., Ahmar, H., Hosseni, H. and Movahed, S. K. (2013). Fabrication of novel redox-active poly (4, 5-dihydro-1, 3-thiazole-2-ylsulfanyl-3-methyl-1, 2-benzediol)-gold nanoparticles film on MWCNTs modified electrode: Application as the electrochemical sensor for the determination of hydrazine. *Sens. Actuators B* 213: 82-91.

[7] Safavi, M. and Tohidi. (2012). Silver paste nanocomposite electrode as a new metallic electrode for amperometric determination of hydrazine. *Anal. Methods,* 4: 2233-2241.

[8] Salimi, L., Miranzadeh, R. and Hallaj. (2008). Amperometric and voltammetric detection of hydrazine using glassy carbon electrodes modified with carbon nanotubes and catechol derivatives. *Talanta* 75: 147-156.

[9] Sardar, C., Czudek, G. and Reynolds. (1996). Dopamine deficits in the brain. *Neuroreport* 7: 910-912.

[10] Umar, Rahman, M. M., Kim, S. H. and Hahn, Y. B. (2008). Zinc oxide nanonail based chemical sensor for hydrazine detection. *Chem. Commun.* 166-169.

[11] Umar, Rahman, M. M. and Shim, Y. B. (2009). ZnO nanorods based hydrazine sensors. *J. Nanosci. Nanotechnol*. 9: 4686-4691.

[12] A. L. Barry. (1980). Procedures for testing antimicrobial agents in agar media. *Antibiotics in laboratory medicines*. Williams and Wilkins Co. Baltimore, USA.

[13] Maybodi, A. S., Ghasemi, S. and Rad, H. G. (2015). A novel sensor based on Ag-loaded zeolite imidazolate framework-8 nanocrystals for efficient electrocatalytic oxidation and trace level detection of hydrazine. *Sens Actuators B* 220: 627-633.

[14] Sarycheva, A. S., Semenova, E. Y. Parshina, E. Y., Brazhe, N. A., Polyakov, A. Y., Kozmenkova, A. Y., Grigorieva, A. V., Maksomov, G. V. and Goodilin, E. A. (2014). Ultrasonic silver rain preparation of SERS substrates. *Mater Lett,* 121: 66-69.

[15] Bauer, A. W., Kirby, W. M. M., Sherris, J. C. and Turck, M. (1966). Antibiotic susceptibility testing by a standard single disc method. *Am J. Clin. Pathol.* 45: 493-496.

[16] Habibi, M. and Azar, P. (2010) Simultaneous determination of ascorbic acid, dopamine and uric acid by use of a MWCNT modified carbon -ceramic electrode and differential pulse voltammetry. *Electrochim. Acta* 55: 5492-5498.

[17] Mari, K. C., Singh, M., Moya, I., Singh, H. and Om, S. (2012). Characterization and photoluminescence properties of some CaO, SrO, and $CaSrO_2$ phosphors co-doped with Eu^{3+} and alkali metal ion s. *Optical Mater*. 34: 1267-1271.

[18] Johnson, P. and Hensley, E. B. (1969). Additive coloration of strontium oxide. *Phys. Rev.* 180: 931.

[19] Roy, J. D., Perkins, T., Kaydanova, D. L., Young, M., Taylor, A., Miedaner, C., Curtis, H-J., Kleebe, D. W. and Reddy, D. S. G. (2008). Preparation and characterization of sol gel derived copper strontium oxide thin films. *Thin Film Solids* 516: 4093-4101.

[20] Theng. (1979). *Formation and properties of clay-polymer complexes,* Elsevier Scientific Pub., Amsterdam.

[21] B-McAuley, C. E., Bnaks, A. O., Siman, T. G. J. and Jones, R. G. C. (2006). The electroanalytical detection of hydrazine: A comparison of the use of palladium nanoparticles supported on boron-dopped diamond and palladium plated BDD microdisc array. *Analyst* 131: 106-110.

[22] Jacobs, M., Peairs, B. and Venton. (2010). Review: Carbon nanotube based electrochemical sensors for biomolecules. *Anal. Chim. Acta* 662: 105-127.

[23] Liu, Z., Liu, R., Peng, Z. and Zhong. (2014). Quasireversible Process of Dopamine on Copper-Nickel Hydroxide Composite/Nitrogen Doped Graphene/Nafion Modified GCE and Its Electrochemical Application. *J. Anal. Methods Chem.* 2014: 1-7.

[24] Hu, C. G., Wang, W. L., Wang, S. X., Zhu, W. and Li, Y. (2003). Investigation on electrochemical properties of carbon nanotubes. *Diamon. Relat. Mater.* 12: 1295-1299.

[25] Feng, W., Luo, J., Zhang, M., Xu, R., Zhang, H., Wu, Y. L., Asiri, A. M., Khan, S. B., Rahman, M. M., Zheng, G. and Zhao, D. (2013). Multi-layered Mesoporous TiO2 Thin films with large pore and highly crystalline frameworks for efficient photoelectrochemical conversion. *J. Mater. Chem.* A 1: 1591-1599.

[26] Vernot, E. H., MacEwen, J. D., Bruner, R. H., Haus, C. C. and Kinkead, E. R. (1985). Long-term inhalation toxicity of hydrazine. *Appl. Toxicol*. 5: 1050-1064.

[27] Correa, G., and B-Martinez, J. (2014). Combustion synthesis process for the rapid preparation of high purity SrO powders. *Mater. Sci. Poland* 32: 682-687.

[28] Jin, Y., Liu, C. M. and Day, S. A. (2007). Little. Enhanced electron emission from functionalized carbon nanotubes with a barium strontium oxide coating produced by magnetron sputter. *Carbon* 45: 587-593.

[29] Salem. (1987). Spectrophotometric and titrimetric determination of catecholamines. *Talanta* 34: 810-812.

[30] Jin, F., Liu, Y. and Day, C. M. (2007). Barium strontium oxide coated carbon nanotubes as field emitters. *Appl. Phys. Lett.* 90: 143114.

[31] Zhang, H., Hattori, K. and Tanabe. (1988). Aldol addition of acetone, catalyzed by solid base catalysts: Magnesium oxide, calcium oxide, strontium oxide, barium oxide, Lanthanum (III) oxide and Zirconium oxide. *Appl. Catal.* 36: 189-197.

[32] Nassef, M., A-E. Radi, A-E. and O'Sullivan, C. K. (2006). Electrocatalytic oxidation of hydrazine at O-aminophenol grafted modified glassy carbon electrode: Reusable hydrazine amprometric sensor. *J. Electroanal. Chem.* 592: 139-146.

[33] Zare, R. and Nasirrizadeh, N. (2007). Heamtoxylin multi-wall carbon nanotubes modified glassy carbon electrode for electrocatalytic oxidation of hydrazine. *Electrochim. Acta* 52: 4153-4160.

[34] Zare, R., Shekari, Z., Nasirizadeh, N. and Jafari, A. A. (2012). Fabrication, electrochemical characteristics and electrcatalytic activity of 4-((2-hydroxyphenylimino)methyl)benzene -1,2-diol electrodeposited on a carbon nanotube modified glassy carbon electrode as a hydrazine sensor. *Catal. Sci. Technol.* 2: 2492-2501.

[35] Razmi, M., Agazadeh, B. and Habibi, A. (2003). Electrocatalytic oxidation of dopamine at aluminum electrode modified with nickel pentacyanonitrosylferrate films, synthesized by electroless procedure. *J. Electroanal. Chem.* 547: 25-33.

[36] Vander, B. (1919). Theory and Operating Characteristics of the Thermionic Amplifier. *Proceedings of the IRE.* 7: 97-128.

[37] Balkhoyor, H. B., Rahman, M. M. and Asiri, A. M. (2016). Effect of Ce-doping into ZnO nanostructures to enhance the phenolic sensor performances. *RSC Adv.* 6: 58236–58246.

[38] Marwani, H. M., Alsafrani, A. E., Asiri, A. M. and Rahman, M. M. (2016). Silica-gel particles loaded with an ionic liquid for separation ofZr(IV) prior to its determination by ICP-OES. *Sensors* 16: 1001.

[39] Hussain, M. M., Rahman, M. M. and Asiri, A. M. (2017). Ultrasensitive and selective 4-aminophenol chemical sensor development based on nickel oxide nanoparticles decorated carbon nanotube nanocomposites for green environment. *J. Environ. Sci.* 53: 27-38.

[40] Cape, A. and Coomes, E-A. (1960). Kinetics of strontium oxide on Tungsten. *J. Chem. Phys.* 32: 210.

[41] Lei, X., Lu, W., Wang, X., Bian, Y., Xue, C., Wang, L. and Li, L. (2012). Fabrication of MnO_2/graphene oxide composite nanosheets and their application in hydrazine detection. *RSC Adv.* 2: 2541-2544.

[42] Li, X. and Lin. (2007). Electrocatalytic oxidation of hydrazine and hydroxylamine at gold nanoparticle-polypyrrole nanowire modified glassy carbon electrode. *Sens. Actuators B* 26: 527-535.

[43] Liu, H., Chen, Z., Lin, J. and Lin, M. (2010). Preparation of Surface Imprinting Polymer Capped Mn-Doped ZnS Quantum Dots and Their Application for Chemiluminescence Detection of 4-Nitrophenol in Tap Water. *Anal. Chem. 82:*7380–7386.

[44] Liu, J., Lie, Y., Jiang, J. and Huang, X. (2010). C@ZnO nanorod array hydrazine electrochemical sensor with improved sensitivity and stability. *Dalton Trans.* 39: 8693-8697.

[45] Xu, J. Q., Han, J, J., Zhang, Y., Sun, Y. A. and Xie, B. (2008). Studies on alcohol sensing mechanism of ZnO based gas sensors. *Sens. Actuators B* 132: 334-339.

[46] Han, J-K., Choi, S-M. and G-H. Lee, G-H. (2007). Synthesis and photocatalytic activity of nanocrytalline TiO_2-SrO composite powders under visible light irradiation. *Mater. Lett.* 61: 3798-3801.

[47] Perry R. and Fuller. (1992). Effect of fluoxetine on serotonin and dopamine concentration in microdialysis fluid from rat striatum. *Life Sci.* 50: 1683-1690.

[48] Wu, J. Fei, S. and Hu. (2003). Simultaneous determination of dopamine and serotonin on a glassy carbon electrode coated with a film of carbon nanotubes. *Anal. Biochem.* 318: 100-106.

[49] Ozoemena, K. I. and Nyokong, T. (2005). Electrocatalytic oxidation and detection of hydrazine at gold electrode modified with iron phthalocyanine complex linked to mercaptopyridine self-assembled monolayer. *Talanta* 67: 162-168.

[50] Amada, K. Y., Asuda, K. Y., Ara, N. F., Siroma, Z., Anaka, H. T., Miyazaki, Y. and Kobayashi, T. (2003). Potential application of anion-exchange membrane for hydrazine fuel cell electrolyte. *Electrochem. Commun.* 5: 892-896.

[51] Kamigaito, O. (1991). What can be improved by nanometer composites ? *J. Japan Soc. Powder Metallurgy* 38.3: 315-321.

[52] Parsons J. and Justice. (1993). Perfusate serotonin increases extracellular dopamine in the nucleus accumbens as measured by in vivo microdialysis. *Brain Res.* 606: 195-199.

[53] Zheng, J. F. and Song. (2009). Curcumin muti-wall carbon nanotubes modified glassy carbon electrode and its electrocatalytic activity towards oxidation of hydrazine. *Sens. Actuators B* 135: 650-655.

[54] Faisal, M., Khan, S. B., Rahman, M. M., Jamal, A., Asiri, A. M. and Abdullah, M. M. (2011). Synthesis, characterizations, photocatalytic and sensing studies of ZnO nanocapsules. *App. Sur. Sci.* 258: 672-677.

[55] Hasnat, S. B., Aoun, S. M. N., Uddin, M. M., Alam, P. P., Koay, S., Amertharaj, Rashed, M. A., Rahman, M. M. and Mohamed, N. (2014). Copper-immobilized Platinum Electrocatalyst for the Effective Reduction of Nitrate in a Low Conductive Medium: mechanism, adsorption thermodynamics and stability. *App. Catal. A: Gen.* 478: 259-266.

[56] Jose-Yacaman, L., Rendon, J., Arenas, M. and Serra, P. (1996). Maya Blue Paint: An Ancient Nanostructured Material. *Sci.* 273: 223-225.

[57] Mohri, M., Tanabe, K. and Hattori, H. (1974). Isomerization of 1-butene catalyzed by strontium oxide. *J. Catal.* 32: 144-147.

[58] Sun, M., Guo, J., Yang, Q., Xiao, N. and Li, Y. (2014). A new fluorescent and colorimetric sensor for hydrazine and its application in biological system. *J. Mater. Chem.* B 2: 1846-1851.

[59] Hasnat, H. A., Islam, M. A., Aoun, S. B., Safwan, J. A., Rahman, M. M. and Asiri, A M. (2015). Composite noble metal films /H^+ conducting solid polymer electrolyte assemblies: Optimization of nitrate reduction activities in the asymmetric sandwich type reactor. *ChemPlusChem* 80: 1634-1641.

[60] Hussain, M. M., Rahman, M. M., Asiri, A. M. and Awual, M. R. (2016). Non-enzymatic simultaneous detection of L-glutamic acid and uric acid using mesoporous Co_3O_4 nanosheets. *RSC Adv.* 6: 80511-80521.

[61] Hussain, M. M., Rahman, M. M. and Asiri, A. M. (2016). Efficient 2-nitrophenol chemical sensor development based on Ce_2O_3 nanoparticles decorated CNT nanocomposites for environmental safety. *PLoS ONE* 11: e0166265.

[62] Hussain, M. M., Rahman, M. M. and Asiri, A. M. (2016). Sensitive L-Leucine sensor based on a glassy carbon electrode modified with SrO nanorods. *Microchim. Acta* 183: 3265-3273.

[63] Hussain, M. M., Rahman, M. M., Arshad, M. N. and Asiri, A. M. (2017). Hg^{2+} sensor Development based on (E)-N′-Nitrobenzylidine-Benzenesulfonohydrazide (NBBSH) Derivatives Fabricated on a Glassy Carbon Electrode with a nafion matrix. *ACS Omega* 2: 420-431.

[64] Rahman, M. M., Jamal, A., Khan, S. B., Faisal. M. and Asiri, A. M. (2012). Fabrication of Highly Sensitive Acetone Sensor Based on Sonochemically Prepared As-grown Ag_2O Nanostructures. *Chem. Engr. J.* 192: 122-128.

[65] Rahman, M. M., Jamal, A., Khan, S. B., Faisal, M. and Asiri, A. M. (2012). Fabrication of methanol chemical sensor based on hydrothermally prepared α-Fe_2O_3 codoped SnO_2 nanocubes. *Talanta* 95: 18-24.

[66] Rahman, M. M., Jamal, A., Khan, S. B., Faisal, M. and Asiri, A. M. (2012). Highly Sensitive Methanol Chemical Sensor Based on Undoped Silver Oxide Nanoparticles Prepared by a Solution Method. *Microchim. Acta* 178: 99-106.

[67] Rahman, M. M., Jamal, A., Khan, S. B. and Faisal, M. (2011). Characterization and Applications of as-grown β-Fe_2O_3 Nanoparticles Prepared by Hydrothermal Method. *J. Nanoparticle Res.* 13: 3789-3799.

[68] Rahman, M. M., Jamal, A., Khan, S. B. and Faisal, M. (2011). Cu-doped ZnO Based Nanostructured Materials for Sensitive Chemical Sensor Applications. *ACS App. Mater. Interfaces* 3: 1346-1351.

[69] Rahman, M. M., Jamal, A., Khan, S. B. and Faisal, M. (2011). Fabrication of Highly Sensitive Ethanol Chemical Sensor Based on Sm-Doped Co_3O_4 Nano-Kernel by Solution Method. *J. Phys. Chem. C* 115: 9503–9510.

[70] Rahman, M. M., Jamal, A., Khan, S. B. and Faisal, M. (2011). Highly Sensitive Ethanol Chemical Sensor Based on Ni-doped SnO_2 Nanostructure Materials. *Biosens. Bioelectron.* 28: 127-134.

[71] Rahman, M. M., Khan, A. and Asiri, A. M. (2015). Chemical sensor development based on poly(o-anisidine)silverized-MWCNT nanocomposites deposited glassy carbon electrodes for environmental remediation. *RSC Adv.* 5: 71370–71378.

[72] Rahman, M. M., Khan, A., Marwani, H. M. and Asiri, A. M. (2016). Hydrazine sensor based on silver nanoparticle-decorated polyaniline tungstophosphate nanocomposite for use in environmental remediation. *Microchim. Acta* 183: 1787-1796.

[73] Rahman, M. M., Umar, A. and Sawada, K. (2009). Development of Amperometric Glucose Biosensor Based on Glucose Oxidase Enzyme Immobilized with Multi-WalledCarbon Nanotubes at Low Potential. *Sens. Actuators B* 137: 327-333.

[74] Rahman, M. M. and Asiri, A. M. (2015). Development of Penicillin G biosensor based on Penicillinase enzymes immobilized biochips. *Biomed. Microdev.* 17: 9.

[75] Rahman, M. M. and Asiri, A. M. (2015). Development of selective and sensitive bicarbonate chemical sensor based on wet-chemically prepared CuO-ZnO Nanorods. *Sens. Actuators* B 214: 82-91.

[76] Rahman, M. M. and Asiri, A. M. (2015). Fabrication of highly sensitive ethanol sensor based on doped nanostructure materials using tiny chips. *RSC Adv.* 5: 63252-63263.

[77] Rahman, M. M. and Asiri, A. M. (2015). One-step electrochemical detection of cholesterol in presence of suitable $K_3Fe(CN)_6$/phosphate buffer mediator by an electrochemical approach. *Talanta* 140: 96-101.

[78] Rahman, M. M. and Asiri, A. M. (2015). Selective choline biosensors based on choline oxidase co-immobilized into self-assembled monolayers on micro-chips at low potential. *Anal. Methods* 7: 9426-9434.

[79] Rahman, M. M., Abu-Zied, B. M., Hasan, M. M., Asiri, A. M. and Hasnat, M. A. (2016). Fabrication of selective 4-aminophenol sensor based on H-ZSM-5 zeolites deposited silver electrodes. *RSC Adv.* 6: 48435–48444.

[80] Rahman, M. M., Balkhoyor, H. B., Asiri, A. M. and Marwani, H. M. (2016). A gold electrode modified with silver oxide nanoparticle decorated carbon nanotubes for electrochemical sensing of dissolved ammonia. *Microchim. Acta* 183: 1677-1685.

[81] Rahman, M. M., Balkhoyor, H. B. and Asiri, A. M. (2017). Phenolic sensor development based on chromium oxide-decorated carbon nanotubes for environmental safety. *J. Environ. Management* 188: 228-237.

[82] Rahman, M. M., Balkhoyor, H. B. and Asiri, A. M. (2016). Ultrasensitive and selective hydrazine sensor development based on Sn/ZnO nanoparticles. *RSC Adv*. 6: 29342-29352.

[83] Rahman, M. M., Ahmed, J., Asiri, A. M. Siddiquey, I. A. and Hasnat, M. A. (2016). Development of ultra-sensitive hydrazine sensor based on facile CoS_2-CNT nanocomposites. *RSC Adv*. 6: 90470-90479.

[84] Rahman, M. M., Ahmed, J. and Asiri, A. M. (2017). A glassy carbon electrode modified with γ-Ce2S3-decorated CNT nanocomposites for uric acid sensor development: a real sample analysis. *RSC Adv.* 7: 14649-14659.

[85] Rahman, M. M., Ahmed, J. and Asiri, A. M. (2017). Development of Creatine sensor based on Antimony-doped Tin oxide (ATO) nanoparticles. *Sens. Actuators B* 242: 167–175.

[86] Rahman, M. M., Hussein, M. A., Alamry, K. A., Al Shehry, F. M. and Asiri, A. M. (2016). Sensitive methanol sensor based on PMMA -G-CNTs nanocomposites deposited onto glassy carbon electrodes. *Talanta* 150: 71–80.

[87] Rahman, M. M., Shiddiky, M. J. K., Rahman, M. A. and Shim, Y. B. (2009). A Lactate Biosensor Based on LDH/NADH Immobilized on a Conducting Polymer/Multiwall Carbon Nanotube Composite Film. *Anal. Biochem.* 384: 159-165.

[88] Rahman, M. M., Hussain, M. M. and Asiri, A. M. (2016). A novel approach towards the hydrazine sensor development by SrO.CNT nanocomposites. *RSC Adv.* 6: 65338–65348.

[89] Rahman, M. M., Hussain, M. M. and Asiri, A. M. (2016). A glutathione biosensor based on a glassy carbon electrode modified with CdO nanoparticles -decorated carbon nanotube in a nafion matrix. *Microchim. Acta* 183: 3255–3263.

[90] Rahman, M. M., Khan, S. B., Asiri, A. M., Marwani, H. M. and Qusti, A. H. (2013). Selective detection of toxic Pb(II) ions based on wet-chemically prepared nanosheets integrated CuO-ZnO nanocomposites. *Composites B Part: Eng.* 54: 215-223.

[91] Rahman, M. M., Khan, S. B., Asiri, A. M., Alamry, K. A., Khan, A. A. P., Khan, A., Rub, M. A. and Azum, N. (2013). Acetone sensor based on solvothermally prepared ZnO doped with Co_3O_4nanorods. *Microchim. Acta* 180: 675-685.

[92] Rahman, M. M., Khan, S. B. and Asiri, A. M. (2015). A microchip based fluoride sensor based on the use of CdO doped ferric oxide nanocubes. *Microchim. Acta* 182: 487-494.

[93] Rahman, M. M., Khan, S. B. and Asiri, A. M. (2013). Chemical sensor development based on polycrystalline gold electrode embedded low-dimensional Ag_2O nanoparticles. *Electrochim. Acta* 112: 422–430.

[94] Rahman, M. M., Khan, S. B. and Asiri, A. M. (2014). Smart methanol sensor based on silver oxide-doped zinc oxide nanoparticles deposited on microchips. *Microchim. Acta* 181: 553–563.

[95] Rahman, M. M., Khan, S. B. and Asiri, A. M. (2014). Fabrication of Smart Chemical sensors Based on Transition-doped-semiconductor Nanostructure Materials with µ-Chips. *PLoS ONE* 9: e85036.

[96] Rahman, M. M., Khan, S. B., Gruner, G., Al-Ghamdi, M. S., Daous, M. A. and Asiri, A. M. (2013). Chloride Ion Sensors based on low-dimensional α-MnO_2-Co_3O_4 Nanoparticles fabricated Glassy carbon electrodes by simple I-V Technique. *Electrochim. Acta* 103: 143-150.

[97] Rahman, M. M., Khan, S. B., Marwani, H. M. and Asiri, A. M. (2015). A SnO_2-Sb2O3 nanocomposite for selective adsorption of lead ions from water samples prior to their determination by ICP-OES. *Microchim. Acta* 182: 579-588.

[98] Rahman, M. M., Khan, S. B., Marwani, H. M. and Asiri, A. M. (2014). Selective Divalent Cobalt Ions Detection Using Ag_2O_3-ZnO Nanocones by ICP-OES Method for Environmental remediation. *PLoS ONE* 9: e114084.

[99] Rahman, M. M., Khan, S. B., Faisal, M., Asiri, A. M. and Alamry, K. A. (2012). Highly Sensitive Formaldehyde Chemical Sensor Based on Hydrothermally Prepared Spinel $ZnFe_2O_4$ Nanorods. *Sens. Actuators B* 171-172: 932-937.

[100] Rahman, M. M., Khan, S. B., Faisal, M., Asiri, A. M. and Tariq, M. A. (2012). Detection of Aprepitant Drug Based on Low-dimensional Un-doped Iron oxide nanoparticles prepared by Solution method. *Electrochim. Acta* 75: 164–170.

[101] Rahman, M. M., Khan, S. B., Faisal, M., Rub, M. A., Al-Youbi, A. O. and Asiri, A. M. (2012). Determination of Olmisartan medoxomil using hydrothermally prepared nanoparticles composed SnO_2-Co_3O_4 nanocubes in tablet dosage forms. *Talanta* 99: 924-931.

[102] Rahman, M. M. (2012). Development of mediator-free acetylcholine sensor co-immobilized with acetylcholine oxidase using micro-chips. *Current Proteomics* 9: 272-279.

[103] Rahman, M. M. (2011). Fabrication of Mediator-free Glutamate Sensors Based on Glutamate Oxidase using Smart Micro-devices. *J. Biomed. Nanotechnol.* 7: 351-357.

[104] Rahman, M. M. (2014). Reusable and mediator-free cholesterol biosensor based on cholesterol Oxidase immobilized onto TGA - SAM modified smart bio-chips. *PLoS ONE* 9: e100327.

[105] Awual, M. R., Alharthi, N. H., Okamoto, Y., Karim, M. R., Halim, M. E., Hasan, M. M., Rahman, M. M. and Islam, M. M. (2017). Ligand field effect for Dysprosium (III) and Lutetium (III) adsorption and EXAFS coordination with novel composite nanomaterials. *Chem. Engr. J.* 320: 427-435.

[106] Lakhkar, J., Neel, E. A. B., Salih, V. and Knowles, J. C. (2009). Strontium oxide doped quaternary glasses: effect on structure, degradation and cytocompatibility. *J. Mater Sci. Mater. Med.* 20: 1339-1346.

[107] Lawrence, R., Deo, J. and Wang. (2004). Electrochemical determination of hydrogen sulfide at carbon nanotube modified electrodes. *Anal. Chim. Acta* 517: 131-137.

[108] Shafi, N., Midgley, J., Watson, D., Smail, G., Strang, R. and MacFarlane, R. (1989). Analysis of biogenic amines in the brain of the American cockroach (Periplaneta Americana) by gas chromatography -negative ion chemical ionisation mass spectrometry. *J. Chromatogr. B: Biomed. Sci. Applications* 490: 9-19.

[109] Ajayan, L., Schadler, P. and Braun. (2003). *Nanocomposite science and technology*. Wiley-VCH, Weinheim.

[110] Griffiths J. De Haseth. (2007). *Fourier transforms infrared spectrometry*. Wiley-Interscience, Hoboken, N. J.

[111] Amlathe, V. K. and Gupta. (1988). Spectrophotometric determination of trace amounts of hydrazine in polluted water. *Analyst* 113: 1481-1483.

[112] Deb, J. W. and Nicholson. (1999). The effect of strontium oxide in glass-ionomer cements. *J. Mater. Sci. Mater. Med.* 10: 471-474.

[113] Hyman, R. and Malenka. (2001). Addiction and the brain: The neurobiology of compulsion and its persistence. *Nature Reviews Neurosci*. 2: 695-703.

[114] Prabakar, S. J. R. and Narayanan, S. S. (2008). Amperometric determination of hydrazine using a surface modified nickel hexacyanoferrate graphite electrode fabricated following a new approach. *J. Electroanal. Chem*. 617: 111-120.

[115] Golabi, S. M. and Zare, H. R. (1999). Electrocatalytic oxidation of hydrazine at a chlorogenic acid (CGA) modified galssy carbon electrode. *J. Electroanl. Chem*. 465: 168-176.

[116] Nikolaus, S., Antke, C. and Muller, H. (2009). In vivo imaging of synaptic functions in the central nervous system: II. Mental and affective disorder s. *Behavioural Brain Res.* 204: 32-66.

[117] Sultana, S., Uddin, R., Khan, M. Z., Umar, K., Ahmed, A. S. and Shahadat, M. (2015). SnO_2eSrO based nanocomposites and their photocatalytic activity for the treatment of organic pollutants. *J. Mol. Struct*. 1098: 393-399.

[118] Khan, S. B., Faisal, M., Rahman, M. M., Abdel-Latif, I. A., Ismail, A. A., Akhtar, K., Al-Hajry, A., Asiri, A. M. and Alamry, K. A. (2013). Highly sensitive and stable phenyl hydrazine chemical sensors based on CuO flower shapes and hollow spheres. *New J. Chem*. 37: 1098-1104.

[119] Khan, S. B., Rahman, M. M., Jang, E. S., Akhtar, K. and Han, H. (2011). Special susceptive aqueous ammonia chemi-sensor Extended applications of novel UV-curable polyurethane -clay nanohybrid. *Talanta* 84: 1005-1010.

[120] Khan, S. B., Rahman, M. M., Marwani, H. M., Asiri, A. M. and Alamry, K. A. (2014). Exploration of silver oxide nanoparticles as a pointer of lanthanum for environmental applications. *J. Taiwan Inst. Chem. Engr.* 45: 2770-2776.

[121] Lee, S. R, Rahman, M. M., Ishida, M. and Sawada, K. (2009). Development of highly sensitive acetylcholine sensor based on acetylcholine by charge transfer techniques esterase using smart biochips. *Trends Anal. Chem.* 28: 196-203.

[122] Lee, S. R, Rahman, M. M., Ishida, M. and Sawada, K. (2009). Fabrication of a Highly Sensitive Penicillin Sensor Based on Charge Transfer Techniques. *Biosens. Bioelectron.* 24: 1877-1882.

[123] Guerra, S. V., Ubota, L. T. K., Xavier, C. R. and Nakagaki, S. (1999). Experimental optimization of hydrazine detection in FIA system using an electrode modified with copper porphyrin zeolite. *Anal. Sci.* 15: 1231-1234.

[124] Skoog, D. A., Hollar, F. J. and Crouch, S. R. (2007). *Principles of Instrumental Analysis,* 6th ed, Belmont, CA: Thomson, 169–173.

[125] Ahmad, I. A., Wani, O. A., Al-Hartomy, A. S., Al-Shihri, A. and Kalam. (2015). Low temperature chemical synthesis and comparative studies of silver oxide nanoparticles. *J. Mol. Struct.* 1084: 9-15.

[126] Seki, T., Akutsu, K. and Hattori, H. (2001). Calcium oxide and Strontium oxide as environmentally benign and highly efficient heterogeneous catalyst for the Tishchenko reaction of furfural. *Chem. Commun.* 11: 1000-1001.

[127] Wang, T. X. and Chen, W. W. (2008). Solid phase preparation of submicron-sized $SrTiO_3$ crystallites from SrO_2 nanoparticles and TiO_2 powders. *Mater. Lett.* 62: 2865-2867.

[128] Yoshitake, T., Kehr, J., Todoroki, Nohta, K. H. and Yamaguchi, M. (2006). Derivatization chemistries for determination of serotonin, norepinephrine and dopamine in brain microdialysis samples by liquid chromatography with fluorescence detection. *Biomed. Chromatogr.* 20: 267-281.

[129] Kim, T. Y., Song, D., Barea, E. M., Lee, J. H., Kim, Y. R., Cho, W., Lee, S., Rahman, M. M., Bisquert, J. and Kang, Y. S. (2016). Origin of high open-circuit voltage in solid state dye-sensitized solar cells employing polymer electrolyte. *Nano Energy* 28: 455-461.

[130] Lu, J., Shu, Z., Wang, N., Haung, W. and Song. (2015). The intrinsic oxidase-like activity of Ag_2O nanoparticles and its application for colorimetric detection of sulfate. *Mater. Lett.* 154: 33-36.

[131] Zhou, Y., Zhou, Y., Liang, X., Feng, H. and Zhou. (2015). Silver nanoparticles on carboxyl -functionalized $Fe_3O_4$with high catalytic activity for 4-nitrophenol reduction. *RSC Adv.* 5: 50505-50511.

[132] Jiang, X. and Lin. (2005). Overoxidized polypyrrole film directed DNA immobilization for construction of electrochemical micro-biosensors and simultaneous determination of serotonin and dopamine. *Anal. Chim. Acta.* 537: 145-151.

[133] Liu, H. and Bai. (2011). Liquid-solid reaction synthesis of $SrTiO_3$ submicron-sized particles. *Mater. Chem. Phys.* 127: 21-23.

[134] Qian, X., Xiong, D., Asiri, A. M., Khan, S. B., Rahman, M. M., Xu, H. and Zhao, D. (2013). A facile route to cage-like mesoporous silica coated ZSM-5 combined with Pt immobilization. *J. Mater. Chem. A* 1: 7525-7532.

[135] Yue, X., Yang, W., Xu, M., Liu, X. and Jia, J. (2015). High performance of electrocatalytic oxidation and determination of hydrazine based on Pt nanoparticles /TiO_2 nanosheets. *Talanta* 144: 1296–1300.

[136] Ding, Y., Wang, L., Zhang, H., Zhang, C. M., Li, Y. and Lei. (2011). Preparation of TiO_2-Pt hybrid nanofibers and their application for sensitive hydrazine detection. *Nanoscale* 3: 1149-1157.

[137] Kim, Y., Bong, S., Kang, Y., Yang, Y., Mahajan, R. and Kim, J. et al., (2010). Electrochemical detection of dopamine in the presence of ascorbic acid using graphene modified electrodes. *Biosens. Bioelectron.* 25: 2366-2369.

[138] Liu, Y., Deng, Y., Sun, Z., Wei, J., Zheng, G., Asiri, A. M., Khan, S. B., Rahman, M. M. and Zhao, D. (2013). Hierarchical Cu_2S Microsponges Constructed from Nanosheets for Efficient Photocatalysis. *Small* 9: 2702-2708.

In: Chemosensory Sensors and Systems ISBN: 978-1-53612-406-4
Editor: Maximilian Brandt

Chapter 3

A REVIEW OF MODERN OLFACTOMETRY, EQUIPMENT AND APPLICATIONS

James E. Hayes[1], Richard M. Stuetz, PhD[1], and Richard J. Stevenson, PhD[2,*],
[1]UNSW Water Research Centre, School of Civil and Environmental Engineering,
University of New South Wales, Sydney, NSW, Australia
[2]Department of Psychology, Macquarie University, Sydney, NSW, Australia

ABSTRACT

The olfactory sense is difficult to measure. Measurement of olfaction covers at least four separate paradigms, all of which require separate methodologies to analyse, and all of which are subject to a litany of known and unknown variables. As a result, devices used to measure olfaction, olfactometers, have progressively developed variations and sophistication. Despite their comparative infancy, olfactometers have been established across a broad spectrum of disciplines; primarily psychology, neuroscience, biology, food science, and environmental

* Communicating author; phone 61-2-98508098; email dick.stevenson@mq.edu.au.

engineering. While these separate disciplines have all shown a need for olfactometer technology, there is also very little cross-disciplinary information sharing. This review is a précis of modern olfactometer techniques across research disciplines in order to collate their growing complexity and utility.

INTRODUCTION

The olfactory sense is difficult to measure [1, 2]. As a result, devices used to measure olfaction, olfactometers, have progressively become more varied and sophisticated. Despite their comparative infancy, olfactometers have been devised for uses across a broad spectrum of disciplines; primarily psychology, neuroscience, biology, food science, and environmental engineering. This range of disciplines illustrates the validity as well as the versatility of olfactometers, and this review is an attempt to classify the multitudinous modern variations.

Historically, the first olfactometers were static, typically designed as a method to have a degree of control over stimulus strength during presentation [3]. Semi-static olfactometers, that is, olfactometers that allowed for varying concentrations, were first designed by Zwaardemaker although Buccola's osmometer (a design that did not control odour exposure) preceded it [1, 3-8]. Dravnieks defines an olfactometer as "an instrument for the preparation and delivery of an odorant- an odour stimulus- to a chemoreceptor system" that "[measures] physiologically, electrophysiologically or psychophysically" the reactions of the subject tested [9]. As the study of olfaction has continued, Dravnieks' definition requires some amendment. With advances in research, it can be added that olfactometers can measure subject's behaviour and attitudes if desired, and that olfactometers often have the extra advantage of recording these measurements without the need of an independent device. An olfactometer was previously considered any instrument that was capable of delivering an odour stimulus to a subject; in this inclusion sniff bottles and cotton wool dipped in odorant could be considered olfactometers [9, 10]. Similarly, "static" olfactometers have fallen out of use in favour of "dynamic" olfactometers [11]. Currently, olfactometers are typically defined as devices created to facilitate stringent demands on the nature of the stimulus; be it a high degree of stimulus reproducibility, wide range of stimulus concentrations, or some other condition that is unobtainable by simpler measures of olfaction [12, 13].

The dynamic elements of an olfactometer vary, but they are based on the demands of testing such as extreme concentration precision of the delivered odour, the dynamic variability of odour concentration, or the delivery of odour sources that is untenable by other methods [13, 14]. To this end, olfactometers offer a much larger range of both measurement techniques as well as measurement qualities. For example, olfactometers can be integrated with functional magnetic resonance imaging (fMRI), they can be used to assess the physiological behaviour of mosquitoes, they can enhance the experience of films, and be used to threshold test environmentally derived complex odorant mixtures [15-18]. No one olfactometer can accomplish all needs, however all olfactometers share several common elements: a dilution source, a method of stimulus production, as well as a method of stimulus delivery [13, 19]. The variations of elements are determined by cost as well as the targeted measure. Cost determines the capacity to control for variables for which there is a universal optimum, for example temperature and humidity stability. The targeted measure will determine how the olfactometer operates on three separate paradigms; the stimulus delivery precision, the stimulus delivery dynamics, as well as the stimulus delivery methodology. To that end olfactometers have variations in their dilution source, stimulus generation, stimulus delivery, and flow control.

DEMANDS OF OLFACTOMETERS

Olfactometers are used in a range of disciplines, including environmental assessment, neuroscience, and psychology. As a result several different measures are implemented. At the most basic, olfactory threshold (OT) is the assessment on whether an individual can detect an odour [1, 11, 20]. Threshold testing can be accomplished in a number of ways, but most often the staircase method is used [21]. This involves the presentation of a series of odour concentrations next to *n* blanks. The series usually starts with the weakest odour concentration, and moves to stronger concentrations as long as the participant makes incorrect guesses. That pattern is reversed when a participant guesses correctly, and their score is usually an average of the concentration levels after the first correct guess [1, 20, 21]. A similar test, the determination of Odour Units (OU; ouE/m^3) in environmental assessment, assesses the number of dilution iterations required for a sampled odour for a panellist to guess it equally as a reference odour concentration (typically *n*-butanol) [22-25]. In this way the determination of ouE/m^3 behaves similarly to a staircase threshold test.

Threshold testing with olfactometers is rare in Psychology; most often other "static" olfactory tests are used due to simplicity and cost [26-28]. In those instances that threshold testing is used, the staircase method is not usually implemented due in part to the rapid concentration changes that can disrupt an olfactometer; as a result other methods such as the method of constant stimuli or method of adjustment are attempted [26, 29, 30]. An important subset of threshold, irritation threshold (the measure of irritancy on the trigeminal nerve) does have some studies using a staircase method olfactometer [29, 31, 32]. Conversely, OU assessment is a vital part of environmental analysis for odourous areas, and has legislative ramifications [17]. Emergent within legislation, and common within Psychological practice, is the measurement of supratheshold values [33, 34].

Suprathreshold olfactory measures include olfactory identification (OI), intensity, discrimination (OD), and hedonic appraisal (OH) [1, 11, 28]. These varying measures do not often translate into a variation of olfactometer, rather, a modification to the program or design for an experiment. In regards to OI and intensity measures, care must be taken to ensure that the olfactometer can allow for a large variation as well as an accurate precision for the flow rate. To this end, olfactometers often must have several iterations of mass flow controllers (MFCs) or rotameters at varying flow limits in order to accommodate for a large flow rate range [3,

9]. Other measures, thanks to most olfactometer's dynamic qualities, can be readily incorporated and recorded.

Neurological studies implementing olfactometers typically have requirements pertaining to the consistency of the odour stream, as well as the capacity for an olfactometer to seamlessly blend or intensify odorants [35-37]. With the addition of a quick time to rise to a particular concentration, neurological olfactometers can be seen to deliver a "square form" of odorant [35, 38]. As a result, neurological olfactometers often require extreme precision and compensatory mechanisms in order to provide consistent airflow [12, 37].

PART 1: DIMENSIONS OF MOST OLFACTOMETERS

Dilution Production

In testing situations, odorants almost always require a degree of dilution as well as a method of moving odorized air towards stimulus delivery. Most often, these tasks are accomplished by the use of a carrier gas which in some methodologies also dilutes the odour. There are a few methods of accomplishing this procedure, modulated predominantly by cost and ease of use. Filtered air derived from an air compressor pump is a relatively inexpensive and popular carrier gas [3, 26, 35]. Compressor pumps can be light and small meaning that portability of the olfactometer is not constrained by the dilution source. Care must be taken to ensure proper filtration of contaminants, and air mixtures will react with some odorants which will change the nature of the odour itself [9]. Somewhat more cumbersome, nitrogen derived from pressurised tanks is completely odourless, and confers far less potential for reactivity with odorous substances. As a result nitrogen is far more useful in systems where odour storage in a gaseous stage is required [32, 38-40]. The need for a constant supply of nitrogen as well as the size of the tanks employed make any olfactometers using pressurised cylinders unfeasible for easy portability. With regards to using cylinders, the use of pressurised tanks need not be limited as being the diluting source; Smeets et al., derived their odorant from a pressurised cylinder which was subsequently metered and diluted with an air stream [29]. A component of testing not yet discussed in olfactometer research is the ecological validity of the carrier gas. It seems possible that a response to an odour will vary if that odour is diluted in a gas that is different to the surrounding environment the odour is typically experienced in. Given that changes in other stimuli cues affect some animals for other senses, it seems likely that some olfactory tests will be affected by carrier gas variation [41]. For this reason, air, when properly filtered, may have experimental advantages over other diluents. This is because it controls for other environmental odours but is also it is a carrier gas that is constantly experienced by most test subjects and so may confer ecological validity.

Less often implemented, odorants are diluted in set amounts of a solution in order to provide varying odour concentrations. Combined with air dilution, these two methods can offer practically any concentration of an odour which can be useful for delivering extremely low concentrations of odorants, such as insect pheromones [9]. However, there are two considerable disadvantages in using diluted solutions. Most obvious, solutions can often interfere with the odour being tested, either by reacting with the odorant, or producing an odour itself [3, 9, 42]. Secondly, the variance of vapour pressure between solvent and odorant affects stimulus strength, and in turn this strength will be further affected by successive testing using the same solution sample [3, 9]. These major disadvantages can be ameliorated by placing solutions on filter paper and allowing the solvent to dissolve, leaving the odorant in a set, small concentration [43].

Stimulus Production

Vapour Saturation

The vapour saturation, or "Dravniek style" design is perhaps the oldest, and most used, modern olfactometer approach [9, 26, 44]. In essence, air passes over a large vessel containing an odorous compound.

This air carries the odour in a calculated concentration, whereupon it is subsequently mixed with pure air to determine a particular odour concentration [44-46]. The vessel must be sufficiently large in order to establish 100% saturation [44]. The degree of saturation is determined by the odorant's diffusion coefficient, air flow rate, environmental conditions, and this may be established by using Fick's diffusion law [3, 9]. Vapour saturation methods allow for fairly quick establishment of a steady state odour stream and subsequent decline once the odour valve has been redirected [47]. Certain specialised designs can reduce the time to establish and decline odour concentration further [47]. The method of vapour saturation is sensitive to changes in temperature and humidity and this may present a problem when trying to compare two odors which may have different conditions required to produce full saturation [9, 48]. Similarly designed, some olfactometers implement odour chambers that contain a sample odorant that air passes around and collects the odour from. These odour chambers and odour sources are unsuitable to establish definite full saturation, however they are beneficial when natural odour sources are used, or when fixed odour concentration is not a concern [49, 50].

Another methodology that most often uses vapour saturation, albeit in a different configuration, is by using a vacuum for air flow. Vacuum methods of drawing the odour are not often used, but possess advantages pertaining to odour control in head space, reduced turbulence, as well as time for odour concentrations to reach steady state [51]. Vacuums draw the odorant through to stimulus delivery, where there is a "break" in the line allowing the participant to sniff through [29]. Negative pressure draws the odour past this break unless the participant is sniffing or odour channels are active, meaning that the odour is not being constantly ejected into headspace [44]. Vacuum pressure can also be used in different ways. Both Laing et al., and Sobel et al., describe vapour saturation olfactometers in where vacuum pressure drew either odourless or saturated air depending on which channel was active. This method permitted very fast, cue-less switching between odor streams which was beneficial to fMRI and odour mixture analyses [29, 44, 48]. The main disadvantages with vacuum-based olfactometers is their complexity of design and subsequent expense [35].

Diffusion

Diffusion methods involve the flow of air across a vessel (Figure 1) which in turn determines the concentration of the odour via measuring by Fick's diffusion law and altering the air flow to adjust [9]. Multiple vessels can be put within a series to increase odour concentration or produce odorant mixtures [12]. Diffusion olfactometers in this way do not always need to use air dilution streams like Dravniek versions, although this limits their application in instances of testing thresholds or other measures that require a variation of odour strength as well as quickly establishing steady state air flow [3, 9]. Teflon and stainless steel seem to be the only appropriate materials for producing the diffusion vessels to reduce contamination of the odour clinging to the sides and changing the odour concentration [3, 9, 12]. Diffusion olfactometers have applications in tasks that require continuous, steady odour concentrations for a comparatively long amount of time, such as fMRI recordings or some types of discrimination tasks [12].

Sealed Bag

When collecting environmentally-derived odours, a useful method is drawing the odours into a bag and transporting them to a specially designed "sealed bag" olfactometer. While the principles of air dilution akin to other styles of olfactometers are still applied in this method, the odourised air delivery is somewhat different. Odour samples are extracted from a sample or environment to be studied and stored in specially designed bags, often made from Mylar, Tedlar, Nalophlan, or an alternative known as Solid-Phase Microextraction (SPME) [17, 40, 52, 53]. These samples are placed in a pressurised container and connected to an outlet that is directed towards the sniffing port or mixed with pure air. As pressure on the bag increases, the odour is pushed measurably through the outlet, whereupon the odour can be manipulated in much the same way as a typical Dravniek style olfactometer [54]. This kind of olfactometer is typically used for environmental assessment; obtaining the Odour Units (OU) of an environmentally-derived sample which may have legislative ramifications [40]. OU is somewhat similar to OT; the sample is diluted in air in a staircase method to be comparable to a panel's detection of a standard odorant, usually *n*-butanol. The OU is subsequently determined by the number of steps in the staircase required to attain this standard threshold value. Determining OU is the primary use of sealed bag olfactometers, but they may also have applications relating to olfactory identification and hedonic appraisal of the environmental odours [17, 33, 55, 56].

Sealed bag olfactometers are useful to derive complex samples, but there are some particular limitations in their application. Because the odour is not derived from a sizable liquid source, the amount of odour is small, meaning that testing must be accomplished promptly, and that continuous sampling for an extended period is fairly unrealistic. In addition, owing mostly to the nature of the odours (that may be partially or wholly unknown) and the demand for several panellists to ensure the veracity of OU results, there is no guarantee that contamination or cross contamination from previous steps or samples does not occur [25]. To remedy this, tubing is often thoroughly flushed between samples and steps.

Despite being the best current technology for taking odour samples, the bags and their handling carry additional risks. Firstly, the type of odour extraction will change the composition of the sample itself [52]. Secondly, the conditions for the sampling and storage of these bags is crucial to fair assessment, as factors such as time, temperature and other variables at these stages will affect the odour collected [25, 40, 52, 57, 58].

Stimulus Delivery

Another consideration in the design of an olfactometer is the kind of port from which the odour is delivered to the test subject. The kind of odorant delivery is largely dependent on what measures are being sought. The vast majority of olfactometers either completely or partially rely on an individual's natural sniffing. At first glance, this might seem unusual as sniffing could be considered an additional, uncontrolled variable between participants. However Laing (1985) established that natural sniffing procured the most accurate results, and for much olfactory research, the natural sniff confers some ecological validity [59]. As a result, "blast olfactometry," that is, olfactometers that artificially force odorised air into a participant's nares, have fallen out of use [3]. Stimulus delivery consist of four main versions; odour ports or cups, face masks, cannulas, or uses of a room for odour delivery [36]. Odour ports and cups are multitudinous, and are used for many different kinds of olfactory assessment. However, the variation between test subjects should be taken into account whenever this kind of delivery system is implemented. Variation in this instance can include the distance between the cup and the participant's nose (which can be controlled by use of a chin rest), as well as any possible variations of where the odourised air is flowing out of the port. Face masks and cannulas do not have these disadvantages, but their lack of easy disentanglement from the participant means that headspace will not be cleared as quickly, meaning that rapid odour testing is unfeasible; typically masks are used for neuroimaging tests [35, 36]. In the case of non-human testing, the odour delivery "room" is often the olfactometer itself, and the participant's behaviour in the space of the olfactometer is what is analysed [28].

PART 2: NON-STANDARD OLFACTOMETER DESIGNS

Aquatic Olfactometers

Research using olfactometers also extends to the analysis of animals that live in aquatic environments. To this end, the design of aquatic olfactometers are often very specialised due in part to the demands of water as a medium throughout the system. Aquatic olfactometers are functionally similar to other non-human olfactometers, but the simultaneous difficulty of the variability of odour concentration control and potential effect on swimming abilities of the organisms tested means that flow is often unused in favour of a static design [62].

"Insect Olfactometers"

Insect olfactometers are olfactometers in which the behaviour of insects (or potentially other small animals) is assessed by their choice of odour streams when placed inside an olfactometer. Typically, there is a choice between two pathways-one that contains the tested odour, and another which contains undiluted air being delivered at an equal flow rate; however there are many exceptions in both methodology and number of pathways [50, 51, 63-67]. Static olfactometers may also be implemented for specialised purposes, such as the investigation of naturally produced chemical messages [64, 68]. For field testing, odour generators have been placed in an environmental setting, and the location of subsequently caught insects is recorded [69]. Using these methods, researchers have been able to determine attractants and repellents for a variety of insects through methods ranging from simple counts to walking path times of the tested creatures [49, 51]. Insect olfactometers have certain constraints with regards to the demands of the participant. Odour concentrations are most often exceptionally dilute compared to human testing levels, and most stimulus delivery methods are unfeasible [43]. Fortunately, test subjects can often be produced within the laboratory to offer unlimited amounts, and the variability between subjects can be reduced by manipulation. Ballhorn and Kautz have suggested that the ecological validity of olfactometer experiments with arthropods may not be as strong as expected, however, further research is required [70].

"Ink-Jet Olfactometers"

Ink-jet olfactometers have very different odour delivery and creation methodologies compared to other types of olfactometer. Within an inkjet olfactometer, a piezoelectric system produces highly precise microdroplets, which have a range of applications not limited to olfactometers [71-73]. However, a design created by Hayes et al., produced an inkjet olfactometer that assessed individuals with neurodegenerative disorders by producing microdroplets that are then heated and delivered to the participant's nares via a small fan [74]. The advantages listed included ease of transport and capacity for precise threshold testing, however more research into the area of inkjet olfactometers is required [74].

Flow Control

Prior to the introduction of MFCs, rotameters/flowmeters dominated olfactometry flow control [3]. Manually controlled flowmeters are inexpensive, as well as being useful in instances where dynamic olfactometry is not important, but rather require steady air flows for a long period of time [16, 47]. Rotameters are sensitive to downstream pressure, which may alter flow value s [60]. MFCs confer several advantages, including very precise measurement, computerised operation, as well as the potential to change air flow very quickly within an experiment [26, 48, 61]. While MFCs allow for practically any desired outcome with regards to flow control, they are expensive and require regular calibration. An alternative, considered more precise than MFCs is the use of needle valves for flow control, for example the DynaScent™ olfactometer [60]. Needle valves are capable of precise adjustment, but are only suitable for small odour concentrations and are not capable of rapid concentration changes; successive needle valves are required for threshold testing. Historically, other methods of flow control were implemented, but within the range of modern olfactometers, alternatives of MFCs, needle valves, or rotameters are rare [3, 9, 10].

Field Olfactometers

As a basis of environmental odour assessment, specially tasked field olfactometers (also known as "scentometers") are taken by panellists to test sites where the environment is sampled directly. Field olfactometers eliminate the disadvantages of sealed bags and ensure naturalistic validity;

temperature, the odorant, and other environmental variables are tested almost instantaneously [75, 76]. This means that the odour is delivered to the panellist in a way that is how it is experienced within the environment. Field olfactometers, in order to be effective, demand well trained panellists, strong methodology, as well as a degree of objectivity [77]. To that end, a demand on the objectivity of detection for most field olfactometers means that it is more susceptible to error than other forms of environmental odour assessment.

Gas Chromatography: Mass Spectrometry-Olfactometry

Gas Chromatography: Mass Spectrometry-Olfactometry (GC-MS/O) is another procedure by which environmental and complex odours are assessed. GC-MS/O involves the odorant separated by GC and then split between an MS sensor and the trained researcher's own nose to record odours in a time-of-flight monitoring [78-80]. The two sets of information can then be cross-tabulated to investigate precise qualities of the odorant's concentration, or simply to assess the separate odorants that produce the compound tested [79, 81, 82]. GC-MS/O is therefore often used to define and identify key components of odorous mixtures. GC-MS/O analysis is affected by variations in the conditions of samples tested [83]. In addition, complex mixtures of odorants are detected by humans as if they were only one compound [44, 84], meaning that the division of an odour's components will mean that synergistic or antagonistic effects of particular odorous compounds within the sample will be lost [85]. For example, Hobbs et al., discovered that a combination of 4-methyl phenol, H2S, acetic acid, and ammonia actually decreased in perception as 4-methyl phenol levels increased [86, 87].

A specialised subgroup of GC-MS/O is GC-PO, or Gas Chromatograph- Pedestal Olfactometers although it is important to note that GC-POs have a variety of naming classifications [88-90]. "Pedestal" refers to the way in which these olfactometers have the ability to produce a constant odour (the "pedestal") which can also then seamlessly add an additional odour [88-91]. GC-PO has an advantage over GC-MS/O with regards to its ability to have combination of odours that aren't necessarily broken into components for analysis which could confer an advantage in regards to ecological validity. Pedestal olfactometers have found specialised use in the analysis of odour mixtures and the food industry [88-91].

Conclusion

Olfactometers have a wide variety of applications as well as variation. Interestingly, while separate disciplines of inquiry have relied on olfactometers, attempts to bridge varying methodologies is rare. This is unfortunate, as different disciplines can offer insight for investigations. For example, despite olfactory threshold testing in psychology and OU testing in environmental assessment being methodologically very similar, there is little between-discipline discussion. As a result, the two disciplines have variations on the most effective sniffing ports, humidity control, as well as panellist test conditions. While olfactometers are undoubtedly multi-faceted in applications, a cross-pollination of techniques between disciplines may produce more effective olfactometers in the future.

References

[1] Doty, R. L., Olfactory system, in Smell and taste in health and disease, T. V. Getchell, et al., Editors. 1991, Raven Press: New York: USA. p. 175- 204.

[2] Press, D. and S. Minta, (2000) The smell of nature: olfaction, knowledge and the environment. *Ethics, Place Environ.* 3(2): p. 173-186.

[3] Prah, J. D., S. B. Sears, and J. C. Walker, Modern Approaches to Air Dilution Olfactometry, in Handbook of Olfaction and Gustation, R.L. Doty, Editor. 1995, Marcel Dekker Inc. New York. p. 227-256.

[4] Zwaardemaker, H., (1889) Compensation von geruchen mittelst des doppelreichmessers. *Fortscher Med.* 7: p. 721- 731.[Compensation of odor by means of the double gauge. *Fortscher. Med.*7: p. 721- 731.]

[5] Zwaardemaker, H., (1888) Over het meten van den reukzin bij het klinisch onderzoek. *Neder Tijdschr Geneesk.* 24: p. 109- 115. [On measuring the smell in a clinical trial. *Neth J Med.* 24: p. 109- 115.]

[6] Doty, R. L. and G. Kobal, Current trends in the measurement of olfactory function, in Handbook of olfaction and gustation, R.L. Doty, Editor. 1995, Marcel Dekker, Inc. New York: USA. p. 191-226.

[7] Berglund, B., et al., (1986) Research equipment for sensory air quality studies of nonindustrial environments. *Environ. Int..* 12: p. 189- 194.

[8] Philpott, C. M., A. Bennett, and G. E. Murty, (2008) A brief history of olfaction and olfactometry. *J Laryngol Otol.* 122(7): p. 657-62.

[9] Dravneiks, A., Instrumental Aspects of Olfactometry, in Methods in Olfactory Research, D. G. Moulton, A. Turk, and J.W. Johnston, Editors. 1975, Academic Press: London. p. 1-62.

[10] Wenzel, B. M., (1948) Techniques in olfactometry; a critical review of the last 100 years. *Psychol Bull.* 45(3): p. 231-247.

[11] Dalton, P. and M. Smeets, Olfactometry: The Human Nose as Detection Instrument, in Handbook of Human Factors and Ergonomics Methods N. Stanton, et al., Editors. 2005, CRC Press: London. p. 1-8.

[12] Johnson, B. N. and N. Sobel, (2007) Methods for building an olfactometer with known concentration outcomes. *J Neurosci Methods.* 160(2): p. 231-45.

[13] Dravnieks, A. and F. Jarke, (1980) Odor Threshold Measurement by Dynamic Olfactometry: Significant Operational Variables. *J Air Pollut Control Assoc.* 30(12): p. 1284-1289.

[14] Bozza, G., C. Calearo, and G. P. Teatini, (1960) On making of a rational olfactometer. *Acta oto-laryng.* 52: p. 189- 209.

[15] Omrani, S.-M., et al., (2010) Fabrication of an olfactometer for mosquito behavioural studies. *J Vector Borne Dis.* 47: p. 17- 25.

[16] Lorig, T., et al., (1999) A computer-controlled olfactometer for fMRI and electrophysiological studies of olfaction. *Behav Modif.* 31(2): p. 370- 375.

[17] Lebrero, R., et al., (2011) Odor Assessment and Management in Wastewater Treatment Plants: A Review. *Crit Rev Env Sci Tech.* 41(10): p. 915-950.

[18] Nakamoto, T. and K. Yoshikawa, (2001) Movie with scents generated by olfactory display using solenoid valves. *IEICE Trans.* E89-A(11): p. 3327- 3332.

[19] Duffee, R. A. and S. S. Cha, (1980) Consideration of Physical Factors in Dynamic Olfactometry. *J Air Pollut Control Assoc.* 30(12): p. 1294-1295.

[20] Wise, P. M., N. Bien, and C. J. Wysocki, (2008) Two Rapid Odor Threshold Methods Compared to a Modified Method of Constant Stimuli. *Chemosense Percept.* 1(1): p. 16-23.

[21] Doty, R. L., et al., (1995) A study of the test-retest reliability of ten olfactory tests. *Chem Senses.* 20(6): p. 645- 656.

[22] Hobbs, P. J., T. H. Misselbrook, and B. F. Pain, (1995) Assessment of odours from livestock wastes by a photoionization detector, an electronic nose, olfactometry and gas chromatography -mass spectrometry. *J agric engng res* 60: p. 137- 144.

[23] Sironi, S., et al., (2007) Odour emission factors for assessment and prediction of Italian rendering plants odour impact. *Chem Eng J.* 131(1-3): p. 225-231.

[24] Sironi, S., et al., (2010) Odour impact assessment by means of dynamic olfactometry, dispersion modelling and social participation. *Atmos Environ*. 44(3): p. 354-360.

[25] Brattoli, M., et al., (2011) Odour detection methods: olfactometry and chemical sensors. *Sensors* (Basel). 11(5): p. 5290-322.

[26] Hayes, J. E., A. L. Jinks, and R. J. Stevenson, (2013) A comparison of sniff bottle staircase and olfactometer-based threshold tests. *Behav Res Methods*. 45(1): p. 178-82.

[27] Schmidt, R. and W. S. Cain, (2010) Making scents: dynamic olfactometry for threshold measurement. *Chem Senses*. 35(2): p. 109-20.

[28] Doty, R. L., Psychophysical measurement of odor perception in humans, in The human sense of smell, D. G. Laing, R. L. Doty, and W. Breipohl, Editors. 1991, Springer: Berlin, Germany. p. 95- 134.

[29] Smeets, M. A., et al., (2007) Odor and irritation thresholds for ammonia: a comparison between static and dynamic olfactometry. *Chem Senses*. 32(1): p. 11-20.

[30] Berglund, B., T. Lindvall, and S. Nordin, (1992) Environmentally Induced Changes in Sensory Sensitivities. *Ann NY Acad Sci*. 641: p. 304-321.

[31] Dalton, P., (2001) Psychophysical Methods in the Study of Olfaction and Respiratory Tract Irritation. *AIHAJ*. 62(6): p. 705-710.

[32] Monse, C., et al., (2010) Use of a calibration gas generator for irritation threshold assessment and as supplement of dynamic dilution olfactometry. *Chem Senses*. 35(6): p. 523-30.

[33] Suffet, I. H., et al., (2009) Sensory Assessment and Characterization of Odor Nuisance Emissions during the Composting of Wastewater Biosolids. *Water Environ Res*. 81(7): p. 670-679.

[34] Sucker, K., et al. (2008) Odor frequency and odor annoyance. Part I: assessment of frequency, intensity and hedonic tone of environmental odors in the field. *Int Arch Occup Environ Health*. 81(6): p. 671-82.

[35] Lundstrom, J.N., et al., (2010) Methods for building an inexpensive computer -controlled olfactometer for temporally-precise experiments. *Int J Psychophysiol*. 78(2): p. 179-89.

[36] Ng, J., et al., (2011) An MR compatible olfactometer for clinical research use. *Proc. Intl. Soc. Mag. Reson. Med.* 19: p. 3659.

[37] Kobal, G., Process for measuring sensory qualities and appartus therefor, U. S. Patent, Editor 1987: US.

[38] de Wijk, R., et al., (1996) An injection olfactometer for humans and a new method for the measurement of the shape of the olfactory pulse. *Behav Res Methods Instrum Comput.* 28(3): p. 383- 391.

[39] Hartell, M. G., et al., Overview of current olfactometer performance across different odorant systems. in Chemistry- and biology- based technologies for contraband detection. 1996. Boston, MA,USA: SPIE.

[40] Muñoz, R., et al., (2010) Monitoring techniques for odour abatement assessment. *Water Res.* 44: p. 5129- 5149.

[41] Delius, J. D., (1992) Categorical discrimination of objects and pictures by pigeons. *Anim Learn Behav*. 20(3): p. 301- 311.

[42] Gamble, K. R. and D. W. Smith, (2009) Discrimination of "odorless" mineral oils alone and as diluents by behaviorally trained mice. *Chem Senses.* 34(7): p. 559-63.

[43] Sharma, K. R. and H. Y. Fadamiro, (2013) Fire ant alarm pheromone and venom alkaloids act in concert to attract parasitic phorid flies, Pseudacteon spp. *J Insect Physiol*. 59(11): p. 1119-24.

[44] Laing, D. G., et al., (1994) Evidence for temporal processing of odor mixtures in humans. *Brain Res.* 615: p. 317- 328.

[45] Dravnieks, A. and W. H. Prokop, (1975) Source Emission Odor Measurement by a Dynamic Forced-Choice Triangle Olfactometer. *J Air Pollut Control Assoc.* 25(1): p. 28-35.

[46] Smeets, M., et al., (2006) Odor and irritation thresholds for ammonia: a comparison between static and dynamic olfactometry. *Chem Senses.* 32(1): p. 11-20.

[47] Walker, J. C., et al., (1990) Apparatus for the automated measurement of the responses of humans to odorants. *Chem Senses.* 15(2): p. 165- 177

[48] Sobel, N., et al., (1997) A method for functional magnetic resonance imaging of olfaction. *J Neurosci Methods.* 78: p. 115- 123.

[49] Walter, A. J., et al., (2010) Walking Response of the Mediterranean Pine Engraver, Orthotomicus erosus, to Novel Plant Odors in a Laboratory Olfactometer. *J Insect Behav.* 23(4): p. 251-267.

[50] Turlings, T. C. J., A. C. Davison, and C. Tamò, (2004) A six-arm olfactometer permitting simultaneous observation of insect attraction and odour trapping. *Physiol Entomol.* 29(1): p. 45-55.

[51] Louise, E. M., et al., (1983) An airflow olfactometer for measuring olfactory responses of hymenopterous parasitoids and other small insects. *Physiol Entomol.* 8(97- 106).

[52] Hudson, N. and G. A. Ayoko, (2008) Odour sampling 1: Physical chemistry considerations. *Bioresour Technol.* 99(10): p. 3982-92.

[53] Bulliner, E., et al., (2006) Characterization of livestock odors using steel plates, solid- phase microextraction, and multidimensional gas chromatography-mass spectrometry-olfactometry. *J Air & Waste Manage Assoc.* 56: p. 1391- 1403.

[54] Choinière, D. and D. Giard, (2013) A new state of the art stationary dynamic dilution olfactometer. *Chem Eng Trans.* 30: p. 217- 222.

[55] Nicolai, R. E., et al., Development of a dynamic olfactometer lab. in 1997 *ASAE Annual International Meeting.* 1997. Minneapolis, Minnesota.

[56] Burlingame, G. A., et al., (2004) Development of an Odour Wheel Classification Scheme of Wastewater. *Wat Sci Technol.* 49(9): p. 201- 209.

[57] Le, H., et al., Stability of volatile sulfer compouds (VSCs) in sample bags- impact of temperature, in 5th IWA Conference on Odours and Air Emissions 2013, International Water Association: San Francisco, California, US.

[58] Laor, Y., et al., (2010) Methodological aspects of sample collection for dynamic olfactometry. *Chem Eng Trans.* 23: p. 55-60.

[59] Laing, D. G., (1985) Optimum performance of odor intensity by humans. *Physiol Behav.* 34: p. 569- 574.

[60] Jaing, J., Development of the next generation dynamic olfactometer- Dynascent, 2003, *Environ Odour* Australia Pty Ltd: Sydney, Australia.

[61] Sezille, C., et al., (2013) A portable experimental apparatus for human olfactory fMRI experiments. *J Neurosci Methods.* 218(1): p. 29-38.

[62] Müller, J. P., et al., (2013) To flee or not to flee: detection, avoidance and attraction of profitable resources by Daphnia magna studied with olfactometer. *J Limnol.* 72(3).

[63] Patt, J. M. and R. S. Pfannenstiel, (2008) Odor-based recognition of nectar in cursorial spiders. *Entomol exp appl.* 127(1): p. 64-71.

[64] Kao, S.-S., F. V. Dunkel, and P. K. Harein, (1984) Behavioral response of the larger black flour beetle (Coleoptera: tenebrionidae) to olfactory cues from food sources. *J Econ Entomol* 77: p. 110- 112.

[65] Giles, D. K., K. M. Heinz, and M. P. Parrella, (1996) Quantitative assessment of insect olfactometer performance by experimental flow analysis. *Biol Control* 44- 47: p. 44- 47.

[66] Blackmer, J. L., C. Rodriguez-Saona, and J. A. Byers, (2004) Behavioural response of lygus hesperus to conspecifics and headspace volatiles of alfalfa in a Y-tube olfactometer. *J Chem Ecol.* 30(8).

[67] Graziosi, I. and L. K. Rieske, (2013) Response of Torymus sinensis, a parasitoid of the gallforming Dryocosmus kuriphilus, to olfactory and visual cues. *Biol Control.* 67(2): p. 137-142.

[68] Rasmann, S., et al., (2005) Recruitment of entomopathogenic nematodes by insect-damaged maize roots. *Nature.* 434(7): p. 732-737.

[69] Torr, S. J., (1990) Dose responses of tsetse flies (Glossina) to carbon dioxide, acetone and octenol in the field. *Physiol Entomol.* 15: p. 93-103.

[70] Ballhorn, D. J. and S. Kautz, (2013) How useful are olfactometer experiments in chemical ecology research? *Commun Integr Biol.* 6(4): p. e24787-1-24787-3.

[71] Grate, J. W., R. G. Ewing, and D. A. Atkinson, (2012) Vapor-generation methods for explosives -detection research. *TrAC Trends in Analytical Chemistry.* 41: p. 1-14.

[72] Bogdan, V. A., et al., Vapor generator for the calibration and test of explosive detectors. in 2008 *IEEE Conference on Technologies for Homeland Security.* 2008. Waltham, MA, USA.

[73] Verkouteren, R. M. and J. R. Verkouteren, (2009) Inkjet metrology: high-accuracy mass measurements of microdroplets produced by a drop-on-demand dispenser. *Anal Chem.* 81: p. 8577- 8584.

[74] Hayes, D. J., et al., A digital olfactometer for smell threshold measurements in neurodegenerative disease diagnostics. in *IEEE Eng. and Med. in Bio. Workshop.* 2007. Dallas, TX.

[75] Newby, B. D. and M. A. McGinley. Ambient odour testing of concentrated animal feeding operations using field and laboratory olfactometers. in *2nd International conference on odour and VOCs: Measurement, regulation, and control techniques.* 2003. Singapore: International Water Association.

[76] McGinley, M. A. and C. M. McGinley. Comparison of field olfactometers in a controlled chamber using hydrogen sulphide as the test odorant. in *2nd International Conference on Odour and VOCs: measurement, regulation, and control techniques.* 2003. Singapore.

[77] Dalton, P., et al., (2011) A Multi-Year Field Olfactometry Study Near a Concentrated Animal Feeding Operation. *J Air Waste Manage Assoc.* 61(12): p. 1398-1408.

[78] Bazemore, R., Off- odor detection utilizing gas chromatography-mass spectrometry /olfactometry (GC-MS/O), 2005, *Volatile Analysis.*

[79] Brattoli, M., et al., (2011) Odour detection methods: olfactometry and chemical sensors. *Sensors.* 11(5): p. 5290-5322.

[80] Zhang, S., et al., (2010) Field air sampling and simultaneous chemical and sensory analysis of livestock odorants with sorbent tubes and GC–MS/olfactometry. *Sensor Actuator B-Chem.* 146(2): p. 427-432.

[81] Brewer, M., S., Overview of odor measurement techniques. (2004) http://www.livestocktrail.uiuc.edu/uploads/sowm/papers/p59-74.pdf (accessed 6th March 2013).

[82] Etiévant, P., et al., (1999) Odor intensity evaluation in gas chromatography-olfactometry by finger span method. *J Agric Food Chem.* 47: p. 1673- 1680.

[83] Kleeberg, K. K., et al., (2005) Development of a simple and sensitive method for the characterization of odorous waste gas emissions by means of solid-phase microextraction (SPME) and GC-MS/olfactometry. *Waste Manag.* 25(9): p. 872-9.

[84] Brambilla, M. and P. Navarotto, (2010) Sensorial analysis of pig barns odour emissions. *Chem Eng Trans*. 23: p. 243- 248.

[85] Brennan, B., (1993) Odour nuisance. *Water Waste Treat.* 36: p. 30-33.

[86] Hobbs, P. J., et al., Relationships between the chemical composition and olfaction of decay odours. in ISEOEN 2000. 2001.

[87] Lebrero, R., et al., (2011) Odor Assessment and Management in Wastewater Treatment Plants: A Review. *Crit Rev Env Sci Tech.* 41(10): p. 915-950.

[88] Williams, R. C., et al., (2009) A Gas Chromatograph-Pedestal Olfactometer (GC-PO) for the study of odor mixtures. *Chem. Percept.* 2: p. 173- 179

[89] Hattori, S., H. Takagaki, and T. Fujimori, (2003) Evaluation of Japanese green tea extract using GC/O with original aroma simultaneously input to the sniffing port method (OASIS). *Food Sci Techn Res.* 9(4): p. 350- 352.

[90] Burseg, K. M. and C. D. De Jong, The Olfactoscan®: in-vivo screening for off-flavour solutions, in *Expression of multidisciplinary flavour science: 12th Weurman symposium.* 2008: Interlaken, Switzerland.

[91] Burseg, K. and C. de Jong, (2009) Application of the Olfactoscan method to study the ability of saturated aldehydes in masking the odor of methional. *J Agric Food Chem.* 57(19): p. 9086-90.

Index

A

B

C

D

E

U

V

X